D1544957

Selection Indices
in
Plant Breeding

Author

R. J. Baker, Ph.D.

Senior Research Scientist
Crop Development Centre
University of Saskatchewan
Saskatoon, Canada

CRC Press, Inc.
Boca Raton, Florida

Library of Congress Cataloging-in-Publication Data

Baker, R. J., 1938-
    Selection indices in plant breeding.

    Bibliography: p.
    Includes index.
    1. Selection (Plant breeding)   I. Title.
SB123.B18   1986   631.5′3   85-29924
ISBN 0-8493-6377-2

Direct all inquiries to CRC Press, Inc., 2000 Corporate Blvd., N.W., Boca Raton, Florida, 33431.

© 1986 by CRC Press, Inc.

International Standard Book Number 0-8493-6377-2

Library of Congress Card Number 85-29924
Printed in the United States

PREFACE

Chapter 1 provides a review of the strategy and methodology of selection indices, as well as of some of the modifications that have been suggested. Recent literature on research relating to the use of selection indices in plant species is reviewed in Chapter 2. This review should provide the reader with insight into current thinking on the subject.

Basic principles of quantitative genetics are reviewed in Chapter 3. These principles are considered in a context which applies particularly to the development of selection indices for plant breeding. This is followed immediately by a detailed consideration in Chapter 4 of the methods that can be used to develop estimates of the genotypic and phenotypic variances and covariances required for selection indices.

The first four chapters provide an overall review of topics which relate to the development of selection indices. Readers who have studied the subject may wish to proceed directly to Chapters 5 and 6 where there is a discussion of some of the issues related to the use of selection indices, and of some related methods that have been recommended for evaluating parental material for breeding programs. Chapter 7 includes an overall evaluation of various modifications that have been proposed for selection indices and their use in plant breeding. This chapter constitutes the author's recommendations as to how selection indices should be used for crop improvement.

Chapters 8 and 9 rely partly on the recommendations developed in Chapter 7 and partly on the methods for estimating variances and covariances that were reviewed in Chapter 4. Chapter 8 constitutes a recipe book for applying selection indices to single or multiple trait improvement in plant breeding programs. In Chapter 9, methods for choosing parents for plant breeding programs are presented.

Chapter 10 and the appendices provide a guide to computer techniques for carrying out the calculations required to develop and apply selection indices.

THE AUTHOR

Dr. R. J. Baker has conducted significant theoretical and applied research on subjects relating to crop improvement for nearly two decades. While employed as a quantitative geneticist at the Agriculture Canada Research Station in Winnipeg, Canada, he worked closely with cereal breeders and pathologists on a range of problems related to the development of superior methods of crop improvement. He currently directs a wheat improvement program at the University of Saskatchewan.

Dr. Baker has published more than 60 papers on selection indices, genotype-environment interaction, breeding methodology in self-pollinated species, methods of genetic analysis, methods of statistical analysis, and on the inheritance and heritability of agronomic and quality traits in cereal crops. He has served as a research advisor in East Africa and is often called upon for statistical advice and advice on computer methodology. He has contributed significantly to graduate student training and to the research efforts of his colleagues.

He received his B.S.A. and M.Sc. from the University of Saskatchewan and his Ph.D. in plant genetics from the University of Minnesota. He spent one year as a Visiting Research Fellow in the Department of Genetics at the University of Birmingham, United Kingdom.

Dr. Baker has been awarded fellowships in the American Society of Agronomy and in the Crop Science Society of America for his research contributions.

ACKNOWLEDGMENTS

Helpful suggestions on content and presentation were contributed by Drs. A. R. Hallauer, R. R. Hill, Jr., D. R. Knott, W. R. Meredith, Jr., and A. E. Slinkard. The author sincerely appreciates the efforts of these scientists who took time from their busy schedules to read parts or all of the draft manuscript. The efforts of Ms. Lori Werdal in typing and proofreading the many equations is also gratefully acknowledged.

The author wishes to acknowledge the encouragement of his wife, Joan, and to thank her for her patience while this book was being written.

To my parents

TABLE OF CONTENTS

Chapter 1

HISTORY OF SELECTION INDEX METHODOLOGY

I. INTRODUCTION

In most crop improvement programs, there is a need to improve more than one trait at a time. Recognition that improvement of one trait may cause improvement or deterioration in associated traits serves to emphasize the need for simultaneous consideration of all traits which are important in a crop species. Selection indices provide one method for improving two or more traits in a breeding program.

This chapter reviews the basic strategy behind the original development of selection indices and some of the modifications that have been developed subsequently. The concepts of optimum selection indices, restricted selection indices, and of other modifications, are introduced. Some of the problems associated with applying these methods to crop improvement are presented.

II. THE BASIC STRATEGY OF OPTIMUM SELECTION INDICES

The use of a selection index in plant breeding was originally proposed by Smith,[1] who acknowledged critical input from Fisher.[2] Subsequently, methods of developing selection indices were modified, subjected to critical evaluation, and compared to other methods of multiple-trait selection.

Today, it is generally recognized that a selection index is a linear function of observable phenotypic values of different traits. The observed value for each trait is weighted by an index coefficient. Symbolically,

$$I = b_1P_1 + \ldots + b_iP_i + \ldots + b_nP_n$$

where P_i represents the observed phenotypic value of the ith trait, and b_i is the weight assigned to that trait in the selection index. With three traits, say, yield, days to maturity, and protein concentration, an index (I) for a particular genotype might be

$$I = 0.3 \times \text{yield} - 1.4 \times \text{days to maturity} + 4.7 \times \text{protein concentration}$$

Then, if the observed yield on a particular genotype was 1800 kg/ha, maturity was 94 days, and protein concentration was 12.4%, the index value for that genotype would be

$$I = 0.3 \times 1800 - 1.4 \times 94 + 4.7 \times 12.4 = 466.68$$

The purpose of using a selection index in plant selection is usually stated as an attempt to select for improved "genotypic worth" of the population. In order to understand what is meant by the term "genotypic worth", one must first have an understanding of what is meant by "genotypic value". If a particular genotype is tested in a large number of environments, its average phenotypic value would be considered to be its genotypic value. Thus, genotypic value is a true value in the sense that it is a measure of the average performance of the genotype within some reference population of environments. As such, genotypic value can rarely be measured, and any estimates of the genotypic value must be by indirect methods (see Chapter 3).

Genotypic worth (W) is defined as a linear function of unobservable genotypic values in which the genotypic value of each trait is weighted by a known relative economic value. In symbols,

$$W = a_1 G_1 + \ldots + a_i G_i + \ldots + a_n G_n$$

where G_i is the unobservable genotypic value of the ith trait and a_i is the relative economic value of that trait. In terms of the example mentioned above, one might consider that an increase of 100 kg/ha in yield has the same economic value as a decrease of 1 day in maturity or an increase of 0.25% in protein concentration. If so, the relative economic values would be 0.01 for an increase of 1 kg/ha in yield, -1.0 for an increase of one day in maturity, and 4.0 for an increase of 1% in protein concentration. Thus, if a particular genotype were known to have genotypic values of 2000 kg/ha for grain yield, 100 days for maturity, and 15% for protein concentration, its genotypic worth would be

$$W = 0.01 \times 2000 - 1.0 \times 100 + 4.0 \times 15 = -20.0$$

If genotypic and phenotypic values have a multivariate normal distribution, there will be a linear regression of genotypic worth on any linear function of phenotypic values. In fact, such a regression may exist with distributions which are not normal. If a linear regression of genotypic worth on phenotypic values does exist, the expected change in genotypic worth due to use of a linear selection index (I) will be

$$\Delta W = (W_s - W_u) = b_{WI}(I_s - I_u) = \frac{\sigma_{WI}}{\sigma_I^2}(I_s - I_u)$$

where ΔW is the response to selection, W_s is the average genotypic worth of the selected sample, W_u is the average genotypic worth of the unselected population, I_s is the average index value of the selected sample, I_u is the average index value for the unselected population, and b_{WI} is the linear regression of genotypic worth on the selection index. If changes in genotypic worth and the selection index are expressed in terms of their respective standard deviations, then

$$\Delta W = (W_s - W_u)\frac{\sigma_W}{\sigma_W} = \frac{\sigma_{WI}}{\sigma_I^2}(I_s - I_u) \text{ and}$$

$$\frac{\Delta W}{\sigma_W} = \frac{(W_s - W_u)}{\sigma_W} = \frac{(I_s - I_u)}{\sigma_I}\frac{\sigma_{WI}}{\sigma_I \sigma_W} = \frac{(I_s - I_u)}{\sigma_I} r_{WI}$$

Rearrangement of the regression equation in this way shows that expected response to index selection, when expressed as a fraction of the standard deviation of genotypic worth, is equal to the product of the standardized selection differential and the coefficient of linear correlation between genotypic worth and the selection index. Since the standardized selection differential will depend only upon the intensity of selection, development of a selection index requires the use of index coefficients that maximize the correlation between genotypic worth and the resulting selection index.

III. OPTIMUM SELECTION INDICES

Smith[1] is usually credited with the initial proposal to use selection indices for the simultaneous improvement of several traits in a breeding program. Smith introduced the concept that the genotypic worth of a plant could be expressed as a linear function of the genotypic values of several traits. In an example, it was considered that an increase of 10 in baking score, or a decrease of 20% in infection by flag smut, was as valuable as an increase of 1 bushel/acre in the yield of wheat, *Triticum aestivum* L. Relative values for increases of one unit in each of the three traits were $1/1 = 1.0$, $1/10 = 0.1$, and $1/-20 = -0.05$. The genotypic worth (W) of a plant or line could therefore be expressed as

$$W = 1.0 \times \text{genotypic value for grain yield}$$
$$+ 0.1 \times \text{genotypic value for baking score}$$
$$- 0.05 \times \text{genotypic value for flag smut infection}$$

Smith,[1] following a suggestion by R. A. Fisher,[2] argued that since it could not be directly evaluated, genotypic worth might best be estimated by a linear function of observable phenotypic values. As indicated previously, the expected change in genotypic worth is

$$\Delta W = (W_s - W_u) = b_{WI} (I_s - I_u)$$

If phenotypic values of all traits in the index are normally distributed, the index itself will be normally distributed. With truncation selection for genotypes with high index scores, the selection differential $(I_s - I_u)$ can be written as $k\sigma_I$, where k is the standardized selection differential and depends only upon the proportion of the population that is selected. In this case, expected response to selection is

$$\Delta W = (W_s - W_u) = k\sigma_I b_{WI}$$

For this reason, Smith[1] argued that index coefficients should be chosen to maximize the product of the standard deviation of the index and the regression of genotypic worth on the index. Since

$$r_{WI} = \frac{\sigma_{WI}}{\sigma_I \sigma_W} = b_{WI} \frac{\sigma_I}{\sigma_W}$$

and since σ_W is constant for any population, maximizing the product of the standard deviation of the selection index and the regression of genotypic worth on the selection index is identical to maximizing the correlation between genotypic worth and the selection index.

Smith[1] showed that the index coefficients which maximized this expression for response to selection were those obtained by solving the set of simultaneous equations which can be expressed in matrix form as **Pb** = **Ga**. For three traits, this set of equations would be written in algebraic form as follows.

$$b_1\sigma^2_{P(1)} + b_2\sigma_{P(12)} + b_3\sigma_{P(13)} = a_1\sigma^2_{G(1)} + a_2\sigma_{G(12)} + a_3\sigma_{G(13)}$$

$$b_1\sigma_{P(12)} + b_2\sigma^2_{P(2)} + b_3\sigma_{P(23)} = a_1\sigma_{G(12)} + a_2\sigma^2_{G(2)} + a_3\sigma_{G(23)}$$

$$b_1\sigma_{P(13)} + b_2\sigma_{P(23)} + b_3\sigma^2_{P(3)} = a_1\sigma_{G(13)} + a_2\sigma_{G(23)} + a_3\sigma^2_{G(3)}$$

In these equations, a_i is the relative economic value of the ith trait, $\sigma^2_{G(i)}$ and $\sigma^2_{P(i)}$ are the genotypic and phenotypic variances of the ith trait, and $\sigma_{G(ij)}$ and $\sigma_{P(ij)}$ are the genotypic and phenotypic covariances between the ith and j^{th} traits. Methods for solving such a set of equations will be discussed and illustrated in Chapter 8.

Table 1
GENOTYPIC AND PHENOTYPIC PARAMETERS FOR THREE TRAITS IN A HYPOTHETICAL POPULATION

| | Trait | | |
Parameter	Yield (kg/ha)	Maturity (days)	Protein (%)
Mean value	3000	100	10
Genotypic variance	100,000	8.0	0.60
Phenotypic variance	250,000	10.0	1.00
Heritability	0.40	0.80	0.60
Relative economic value	0.01	−1.0	4.0

	Pairs of traits		
	Yield-maturity	Yield-protein	Maturity-protein
Genotypic correlation	0.3	−0.5	0.0
Genotypic covariance	268.3	−122.7	0.0
Phenotypic correlation	0.4	−0.6	−0.2
Phenotypic covariance	632.4	−300.0	−0.63

The nature of these equations might be better understood by reference to a set of hypothetical data for three traits (Table 1). Based on the parameters given in Table 1, the following three equations could be solved to give the index coefficients required for simultaneous improvement of grain yield, maturity, and protein concentration.

$$250{,}000 \; b_1 + 632.4 \; b_2 - 300.0 \; b_3 = 0.01(100{,}000) - 1.0(268.3) + 4.0(-122.47) = 241.820$$

$$632.4 \; b_1 + 10.0 \; b_2 - 0.63 \; b_3 = 0.01(268.3) - 1.0(8.0) + 4.0(0.0) = -5.317$$

$$-300.0 \; b_1 - 0.63 \; b_2 + 1.00 \; b_3 = 0.01(-122.47) - 1.0(0.0) + 4.0(0.60) = 1.175 \qquad (1)$$

The estimation of index coefficients for an optimum selection index will require knowledge of the relative economic values of traits as well as the genotypic and phenotypic variances and covariance among these traits. It will be seen that a major obstacle to the use of selection indices is the effort required to obtain suitably precise estimates of the required variances and covariances.

In some cases, relative economic values of some traits included in the index may be set to zero. These secondary traits are not of any direct economic importance, but may serve to enhance overall response in genotypic worth. In these cases, the right hand sides of the index equations need to be modified by dropping any terms for which the corresponding economic value has been set to zero. Furthermore, estimates of genotypic variances or covariances of these particular traits will no longer be required to develop optimum index coefficients. In fact, Smith[1] considered an example where genotypic worth was equal to the genotypic value of grain yield alone, but where the selection index was to include information about ear number per plant, average number of grains per ear, average weight per grain, and weight of straw, all expressed on logarithmic scales.

The Smith[1] method of developing selection indices will be referred to as an "optimum" index in the sense used by Williams.[3] This is to differentiate this method from more recent modifications to the original proposal. If genotypic and phenotypic values are distributed with a multivariate normal distribution, the index coefficients calculated from population parameters in the way proposed by Smith[1] will give maximum response in genotypic worth.

In practice, however, one will use estimates of the corresponding population parameters and the resulting index coefficients will be estimates of the optimal index coefficients. Williams[3] referred to these indices as "estimated" indices.

Hazel[4] took a slightly different, but equivalent, approach to the development of selection indices. The equation for expected change in genotypic worth can be modified in the following way.

$$\Delta W = (W_s - W_u) = b_{WI} (I_s - I_u) = b_{WI} k \sigma_I = k \frac{\sigma_{WI}}{\sigma_I} = k \, r_{WI} \, \sigma_W$$

Since σ_W will be constant for a given population and a given set of relative economic weights, the maximum response to index selection will be achieved if the correlation between genotypic worth and the selection index is maximized. Hazel[4] used this relationship as the basis for choosing index coefficients.

The equations given by Hazel[4] are equivalent to those given by Smith.[1] Dividing both sides of the ith equation by $\sigma_W \sigma_{P(i)}$, the equations of Smith can, for three traits, be written as follows.

$$b_1 \frac{\sigma_{P(1)}}{\sigma_W} + b_2 r_{P(12)} \frac{\sigma_{P(2)}}{\sigma_W} + b_3 r_{P(13)} \frac{\sigma_{P(3)}}{\sigma_W} = r_{P(1)W}$$

$$b_1 r_{P(12)} \frac{\sigma_{P(1)}}{\sigma_W} + b_2 \frac{\sigma_{P(2)}}{\sigma_W} + b_3 r_{P(23)} \frac{\sigma_{P(3)}}{\sigma_W} = r_{P(2)W}$$

$$b_1 r_{P(13)} \frac{\sigma_{P(1)}}{\sigma_W} + b_2 r_{P(23)} \frac{\sigma_{P(2)}}{\sigma_W} + b_3 \frac{\sigma_{P(3)}}{\sigma_W} = r_{P(3)W}$$

Hazel[4] used the method of path coefficients to show that the right hand side of the ith equation can be expressed as

$$r_{P(i)W} = h_i \sum a_j r_{G(ij)} \frac{\sigma_{G(j)}}{\sigma_W}$$

where summation is over the subscript j of the economically important traits, and h_i is the square root of the heritability of the ith trait.

Using the parameters from Table 1, the variance of genotypic worth is given by

$$\sigma_W^2 = (0.01)^2 (100,000) + (-1.0)^2 (8.0) + (4.0)^2 (0.60)$$

$$+ (2)(0.01)(-1.0)(268.3) + (2)(0.01)(4.0)(-122.47)$$

$$+ (2)(-1.0)(4.0)(0.0) = 12.4364$$

Thus, $\sigma_W = (12.4364)^{0.5} = 3.5265$. If Equations 1 are each divided by the corresponding phenotypic standard deviation and by σ_W, they take the following form.

$$141.7825 \, b_1 + 0.3586 \, b_2 - 0.1701 \, b_3 = \quad 0.1371$$

$$57.7080 \, b_1 + 0.8967 \, b_2 - 0.0565 \, b_3 = -0.4768$$

$$-85.0695 \, b_1 - 0.1786 \, b_2 + 0.2836 \, b_3 = \quad 0.3332$$

Since the Hazel[4] equations can be derived directly from those of Smith[1] by dividing each side of each equation by the same constant, both sets of equations will give the same estimates for the index coefficients.

From Hazel's[4] terminology, one can see that, in calculating optimum index coefficients, one must have knowledge of (1) the genotypic variance of each trait, (2) the genotypic correlations between each pair of traits, (3) the heritability of each trait, (4) the phenotypic variances of each trait, and (5) the phenotypic correlations between each pair of traits. Hazel's equations emphasize the point that the development of optimum selection indices requires estimation of numerous population parameters.

Hazel[4] developed selection indices for the simultaneous improvement of three traits in swine, *Sus domesticus,* based on (1) two of the three traits measured on juvenile animals, (2) those same two traits plus a third trait measured on the animal's mother, and (3) those same three traits plus the average of the first two for the litter in which the animal was born. These examples serve to show that the traits considered important in determining genotypic worth need not all be included in the index. Furthermore, the index may include traits other than those specified in the definition of genotypic worth, as well as traits measured on relatives of those undergoing selection.

In developing selection indices, Hazel[4] cautioned that the indices developed for one herd may not be widely applicable. Relative economic values may differ from locality to locality or with the nature of the operation. Genotypic variances and covariances may differ from herd to herd and phenotypic variances and covariances may differ because of different management practices. Moreover, initial estimates of population parameters may be subject to large estimation errors.

For genotypic worth defined as the genotypic value of a single trait, say trait 1, the Smith[1] equations, for a three trait index, are as follows.

$$b_1\sigma_{P(1)}^2 + b_2\sigma_{P(12)} + b_3\sigma_{P(13)} = \sigma_{G(1)}^2$$

$$b_1\sigma_{P(12)} + b_2\sigma_{P(2)}^2 + b_3\sigma_{P(23)} = \sigma_{G(12)}$$

$$b_1\sigma_{P(13)} + b_2\sigma_{P(23)} + b_3\sigma_{P(3)}^2 = \sigma_{G(13)}$$

In a similar way, the equations of Hazel[4] simplify to

$$b_1\frac{\sigma_{P(1)}}{\sigma_{G(1)}} + b_2 r_{P(12)}\frac{\sigma_{P(2)}}{\sigma_{G(1)}} + b_3 r_{P(13)}\frac{\sigma_{P(3)}}{\sigma_{G(1)}} = h_1$$

$$b_1 r_{P(12)}\frac{\sigma_{P(1)}}{\sigma_{G(1)}} + b_2\frac{\sigma_{P(2)}}{\sigma_{G(1)}} + b_3 r_{P(23)}\frac{\sigma_{P(3)}}{\sigma_{G(1)}} = h_2$$

$$b_1 r_{P(13)}\frac{\sigma_{P(1)}}{\sigma_{G(1)}} + b_2 r_{P(23)}\frac{\sigma_{P(2)}}{\sigma_{G(1)}} + b_3\frac{\sigma_{P(3)}}{\sigma_{G(1)}} = h_3$$

If, for the population characteristics given in Table 1, the relative economic values of days to maturity and protein concentration were set to zero, the equations for developing a selection index to improve yield alone would be as follows.

$$250,000\ b_1 + 632.4\ b_2 - 300.0\ b_3 = 100,000$$

$$632.4\ b_1 + 10.0\ b_2 - 0.63\ b_3 = 268.3$$

$$-300.0\ b_1 - 0.63\ b_2 + 1.00\ b_3 = -122.47$$

If several traits have economic importance, Henderson[5] has shown that the optimum selection index for total genotypic worth can be obtained by first developing indices, I_i, for each trait and then weighting each index by the economic value of the corresponding trait. Thus, if I_i is the selection index for the ith trait, then

$$I = a_1 I_1 + \ldots + a_i I_i + \ldots + a_n I_n$$

is the optimum index for improving overall genotypic worth. The advantage of this approach is that changes in relative economic values will not require calculation of new selection indices for each trait. All that will be required is a reweighting of the original indices.

For the example considered above, one would estimate index coefficients for an index, I_Y, for increasing yield, for an index, I_M, for increasing days to maturity, and for an index, I_P, for increasing protein concentration. All three indices would be linear functions of yield, days to maturity, and protein concentration. Then, an index for the simultaneous improvement of all three traits would be

$$I = 0.01 \, I_Y - 1.0 \, I_M + 4.0 \, I_P$$

IV. RELATIVE EFFICIENCY OF INDEX SELECTION, INDEPENDENT CULLING, AND TANDEM SELECTION

Three methods of selection, which are recognized as appropriate for simultaneous improvement of two or more traits in a breeding program, are index selection, independent culling, and tandem selection. A selection index is a single score which reflects the merits and demerits of various traits. Selection among genotypes is based on the relative values of the index scores. If selection is conducted for several generations, one might consider using the same index in each generation.

Independent culling requires the establishment of minimum levels of merit for each trait. An individual with a phenotypic value below the critical culling level for any trait is removed from the population. With tandem selection, one trait is selected until it is improved to a satisfactory level. Then, in the next generation, selection for a second trait is practiced within the selected population, and so on for the third and subsequent traits. Differences among the three basic approaches are demonstrated diagrammatically in Figure 1.

Hazel and Lush[6] investigated the relative efficiencies of these three selection methods. They considered expected response to truncation selection when phenotypic values are normally distributed, and when both genotypic and environmental correlations between traits are zero. For index selection, expected response in overall genotypic worth is

$$R_1 = k \, [a_1^2 h_1^4 \sigma_{P(1)}^2 + \ldots + a_n^2 h_n^4 \sigma_{P(n)}^2]^{0.5}$$

where k is the standardized selection differential, σ_P is the phenotypic standard deviation, h^2 is the heritability, and the subscripts refer to the traits included in the index.

For independent culling of uncorrelated traits, response in overall genotypic worth is expected to be

$$R_2 = a_1 h_1^2 k_1 \sigma_{P(1)} + \ldots + a_n h_n^2 k_n \sigma_{P(n)}$$

where k_i is the standardized selection differential corresponding to the fraction saved when culling for the ith trait. For single trait (tandem) selection, expected response is

$$R_3 = k \, \sigma_P \, h^2$$

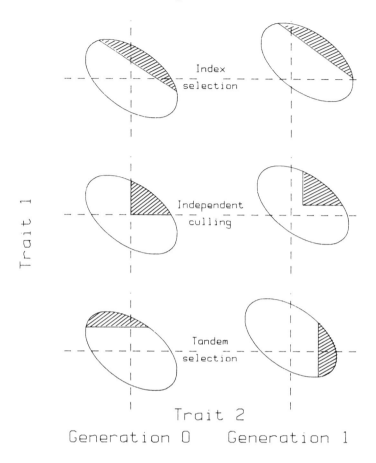

FIGURE 1.   A diagrammatic comparison of index selection, independent culling, and tandem selection for two negatively correlated traits.

To further facilitate comparison of the three selection methods, Hazel and Lush[6] considered only those cases for which the product of relative economic value, heritability, and phenotypic standard deviation, was the same for all traits. Under these more restrictive conditions, expected responses to one cycle of selection for each method can be expressed as follows.

Index: $\qquad\qquad R_1 = n^{0.5}\,k\,a\,h^2\,\sigma_P$

Independent culling: $\quad R_2 = n\,k_3\,a\,h^2\,\sigma_P$

Tandem: $\qquad\qquad R_3 = k\,a\,h^2\,\sigma_P$

In these equations, k depends upon the fraction p of the population selected by tandem or index selection while k_3 depends upon culling of a fraction $p^{1/n}$ for each of the n traits. Under these very specific conditions, index selection will be $n^{0.5}$ times as efficient as tandem selection, regardless of selection intensity. Hazel and Lush presented a geometric argument as to why this should be so.

These equations also show that independent culling for uncorrelated traits is always intermediate in efficiency to index and tandem selection. Independent culling is expected to approach the efficiency of index selection as the number of traits decreases, and as the selected portion of the population becomes smaller (more intense selection). The advantage of independent culling over tandem selection will decrease as the number of traits increases and when selection becomes less intense. Although Hazel and Lush[6] found it necessary to use several restrictive assumptions to compare efficiencies of the three methods of multitrait selection, their results provided a strong argument for the use of selection indices in breeding programs.

The simplified equations of Hazel and Lush[6] cannot be applied to the hypothetical population described in Table 1. First, the genotypic and phenotypic correlations between each pair of the three traits would have to be zero. Then, it would be necessary for the relative economic weights to be changed to -0.79 and 3.33 for days to mature and protein concentration, respectively, if the traits were to be considered equally important in the sense specified by Hazel and Lush.

Young[7] also evaluated these selection methods, but with fewer restrictive assumptions, and came to similar conclusions. Young first considered relative responses to the three selection methods when traits were independent, but not of equal importance. Under these conditions, index selection is more efficient than tandem selection, with the advantage becoming greater as the number of traits increases and as the relative importances of the traits approach a common value.

In comparing index selection with independent culling, Young[7] first estimated the optimum culling levels for each of the uncorrelated traits. The superiority of index selection over independent culling was greatest when the traits were of equal importance and when selection intensity was low to intermediate.

In considering relative efficiencies of selection for correlated traits, Young[7] limited discussion to two traits. Both genotypic and phenotypic correlations have important effects on the relative efficiencies of the three selection methods. Generalization of the results is difficult. However, it does seem that selection using an optimum index will always be at least as efficient as selection based on optimum culling levels. Similarly, selection using optimum culling levels should always be at least as efficient as using tandem selection.

Finney[8] suggested more generalized formulas for studying the relative efficiencies of these three selection methods. Central to Finney's proposal was the concept that covariances between genotypic values of individual traits and overall genotypic worth play a key role in determining the relative effectiveness of the three methods. Expected change in genotypic worth in response to selection for a single trait is proportional to the ratio of the covariance (between genotypic value for that trait and the total genotypic worth) to the phenotypic standard deviation for that trait. If one of these ratios is far larger than any other, tandem selection for that trait, or independent culling with high selection intensity for that trait, should give responses in overall genotypic worth which are close to that expected from index selection. Finney suggested that such an approach would have allowed Young[7] to develop results with wider applicability. For the traits listed in Table 1, the ratio of the covariance between genotypic value and genotypic worth to the phenotypic standard deviation is 0.48 for yield, -1.68 for days to maturity, and 1.17 for protein concentration. Since these three values are similar, it is likely that index selection would be considerably more effective than independent culling or tandem selection for improving genotypic worth.

Pesek and Baker[9] used computer simulation to compare tandem and index selection in advanced generations of inbred species. They considered selection among random F_6 plants from a cross between two homozygous parents, and then among F_7 progeny of the selected plants. For tandem selection, equal proportions were selected for one trait in the F_6 and for the other in the F_7. A selection index was calculated from the F_6 data and used to select in both generations. Their simulations were based on a 10 locus genetic model and included only negative genotypic correlations varying from -0.95 to 0.0, heritabilities from 0.3 to 0.8, and ratios of relative economic importance of 0.5 to 1.5. Their results indicated that index selection should be considerably more efficient than tandem selection for improving genotypic worth for two negatively correlated traits.

In some applications, a selection index based on several traits is used to improve a single trait. It is not difficult to show that, if the index coefficients are reliable estimates of the optimum coefficients, and if the trait included in the definition of genotypic worth is included in the index, selection based on the index will always give greater response than direct selection for that trait by itself. When index coefficients are chosen to maximize the correlation between an index and the genotypic value of one trait, the correlation will always be greater than the correlation with the phenotypic value of that trait, provided that trait is included in the index.

V. RESTRICTED SELECTION INDICES

Kempthorne and Nordskog[10] considered that breeders may sometimes wish to maximize genotypic worth while restricting change in some traits. As an example, they described a case where a poultry breeder wished to maximize economic value based on egg weight, body weight, egg production, and fraction of blood spots, while maintaining body weight at an intermediate level. They calculated such an index. The index was equal to the unrestricted index multiplied by a modifying matrix. In general, imposing a restriction on changes in any trait will reduce the correlation between genotypic worth and the resulting index, and will therefore reduce expected response to selection. The extent of the reduction will depend on the particular correlational structure of the population.

Tallis[11] provided a further extension of the restricted selection index. Rather than restricting response to a fixed value, Tallis developed a method for restricting response to a proportion of a fixed value. If the value is specified as zero, the Tallis index is the same as that of Kempthorne and Nordskog.[10] Harville[12] described such indices as having proportionality constraints. This terminology is more suitable than the "optimum genotype" terminology used by Tallis.[11] Harville[12] considered the case where selection is meant to improve genotypic values of different traits by specified amounts.

Pesek and Baker,[13] in considering the problem of assigning relative economic weights, proposed that selection indices be developed by specifying "desired genetic gains" for each trait. They developed a method for calculating a selection index that would move the genotypic means in the desired direction. Their method appears to be a specific case of the proportional constraint method developed by Tallis.[11] While Pesek and Baker[14] did refer to Tallis'[11] paper, they apparently did not realize that their method of "desired gains" was identical to that of Tallis when proportionality constraints are attached to all traits in the index. Similarly, a more recent paper by Yamada et al.[15] also appears to be identical to that of Tallis.[11]

Tai[16] developed a generalization of the Pesek and Baker index that could be applied when some of the traits in the index are not constrained. It appears that the modification proposed by Tai was covered in an earlier paper by Harville.[17] The indices of Harville,[17] Tai,[16] and Pesek and Baker[14] are all equivalent to that of Tallis[11] if constraints are specified for all the traits in the index. If constraints are not applied to all traits, then it appears that the Tallis[11] index is not as efficient as that developed independently by Harville[17] and by Tai.[16]

VI. GENERAL SELECTION INDICES

Hanson and Johnson[18] noted that any selection index is calculated from a specific population of genotypes tested in specific environments and questioned whether or not such an index could be used as a general index in a breeding program. They suggested it would be more efficient to combine information from a series of experiments to obtain an average or general index. In averaging data from several experiments, the purpose would be to develop an index which would maximize response in reference to all possible genotypes evaluated over all possible environments. They argued that an adequate criterion would be to choose index coefficients so that average genotypic worth would be maximized. Since an exact solution is impractical, they recommended that index coefficients be estimated from pooled genotypic and phenotypic variance-covariance matrices, and that the average undergo a correction to give final estimates. Hanson and Johnson provide a worked example for two populations of soybean, *Glycine max*. Caldwell and Weber[19] also discussed the possibility of developing general selection indices for use in crop breeding.

VII. OTHER MODIFIED SELECTION INDICES

Elston[20] considered the problem of ranking individuals based on their phenotypic values, and proposed a multiplicative index for this purpose. In the Elston index, data for each trait are transformed so that high values are desirable and distributions are similar, at least in number of modes. The index is then constructed by subtracting the minimum sample value for each trait and forming the product of the adjusted values. This procedure is based strictly on phenotypic values and does not require estimation of genotypic and phenotypic parameters. In a sense, the index assumes that each trait receives equal emphasis in the selection or ranking process.

In assessing the use of selection indices, Williams[3] evaluated the use of a base index. A base index was defined as one in which the relative economic values are used as the index coefficients. The index is similar to that of Elston[20] in that it does not require estimates of genotypic and phenotypic parameters. Williams[3] noted that the optimum index of Smith[1] adjusts the weighting coefficients in an attempt to correct for differences in genotypic variability, as well as for correlations between traits. The base index should approach the optimum index in efficiency if correlations between traits are small, and if the variability for each trait reflects its relative importance.

VIII. SUMMARY

The optimum selection index for improving a specified linear function of genotypic values is a linear function of phenotypic values in which the weights attached to each phenotypic value are chosen to maximize the correlation between genotypic worth and the selection index. Estimation of optimum index coefficients requires knowledge of genotypic and phenotypic variances and covariances among traits, or, equivalently, of heritabilities and of genotypic and phenotypic correlations among traits. An index for a case where relative economic values are nonzero for more than one trait can be calculated by first calculating an index for each trait and then weighting the individual indices by the relative economic values of the corresponding traits.

For simultaneous improvement of genotypic worth of several traits, index selection should always be at least as effective as independent culling which, in turn, should always be at least as effective as tandem selection. The advantage of using an optimum index will be greater if the relative economic values are equal, and if genotypic correlations between traits are low or negative.

There are instances where a plant breeder may wish to improve genotypic worth while restricting change in one or more traits to some particular value, or at least to some proportion of a specified value. Theory has been developed for estimating indices which meet the requirements of fixed or proportional constraints. It is apparent that people working in plant breeding have not been fully aware of these theoretical developments. More recent papers on restricted selection indices suggest that earlier proposals might not result in optimum response.

Indices based on average estimates of genotypic and phenotypic parameters, while never having as great a predicted response as specific selection indices, often show realized responses that are in close proximity to those obtained by using specific indices. Such generalized indices may give the best average response.

Other modifications to the selection index methodology include the use of a base index, where relative economic values are used for index coefficients, and a weight-free index based solely on observed phenotypic values.

REFERENCES

1. **Smith, H. F.**, A discriminant function for plant selection, *Ann. Eugenics,* 7, 240, 1936.
2. **Fisher, R. A.**, The use of multiple measurements in taxonomic problems, *Ann. Eugenics,* 7, 179, 1936.
3. **Williams, J. S.**, The evaluation of a selection index, *Biometrics,* 18, 375, 1962.
4. **Hazel, L. N.**, The genetic basis for constructing selection indexes, *Genetics,* 28, 476, 1943.
5. **Henderson, C. R.**, Selection index and expected genetic advance, in Statistical Genetics and Plant Breeding, Publ. No. 982, Hanson, W. D. and Robinson, H. F., Eds., National Academy of Sciences, Natioal Research Council, Washington, D.C., 1963, 141.
6. **Hazel, L. N. and Lush, J. L.**, The efficiency of three methods of selection, *J. Hered.,* 33, 393, 1942.
7. **Young, S. S. Y.**, A further examination of the relative efficiency of three methods of selection for genetic gains under less restricted conditions, *Genet. Res.,* 2, 106, 1961.
8. **Finney, D. J.**, Genetic gains under three methods of selection, *Genet. Res.,* 3, 417, 1962.
9. **Pesek, J. and Baker, R. J.**, Comparison of tandem and index selection in the modified pedigree method of breeding self-pollinated species, *Can. J. Plant Sci.,* 49, 773, 1969.
10. **Kempthorne, O. and Nordskog, A. W.**, Restricted selection indices, *Biometrics,* 15, 10, 1959.
11. **Tallis, G. M.**, A selection index for optimum genotype, *Biometrics,* 18, 120, 1962.
12. **Harville, D. A.**, Index selection with proportionality constraints, *Biometrics,* 31, 223, 1975.
13. **Pesek, J. and Baker, R. J.**, Desired improvement in relation to selection indices, *Can. J. Plant Sci.,* 49, 803, 1969.
14. **Pesek, J. and Baker, R. J.**, An application of index selection to the improvement of self-pollinated crops, *Can. J. Plant Sci.,* 50, 267, 1970.
15. **Yamada, Y., Yokouchi, K., and Nishida, A.**, Selection index when genetic gains of individual traits are of primary concern, *Jpn. J. Genet.,* 50, 33, 1975.
16. **Tai, G. C. C.**, Index selection with desired gains, *Crop Sci.,* 17, 182, 1977.
17. **Harville, D. A.**, Optimal procedures for some constrained selection problems, *J. Amer. Stat. Assoc.,* 69, 446, 1974.
18. **Hanson, W. D. and Johnson, H. W.**, Methods for calculating and evaluating a general selection index obtained by pooling information from two or more experiments, *Genetics,* 42, 421, 1957.
19. **Caldwell, B. E. and Weber, C. R.**, General, average and specific selection indices for yield in F_4 and F_5 soybean populations, *Crop Sci.,* 5, 223, 1965.
20. **Elston, R. C.**, A weight-free index for the purpose of ranking or selection with respect to several traits at a time, *Biometrics,* 19, 85, 1963.

Chapter 2

THE APPLICATION OF SELECTION INDICES IN CROP IMPROVEMENT

I. INTRODUCTION

Much of the recent literature on the use of selection indices for crop improvement concerns self-pollinated crops or maize, *Zea mays*. This reflects the amount of research being conducted with these species rather than the effectiveness of selection indices in their improvement.

The various applications of selection indices can be divided into two major groups. The first group includes those instances where the objective is to improve a single quantitative trait. In this group, selection indices may incorporate data from related traits, or from the same trait in related plants, to increase the effectiveness of selection.

The second group of applications includes those instances where the objective is to improve two or more traits simultaneously. In this group, significant problems arise in assigning relative economic weights to the various traits. Because of this problem, researchers have developed modified methods, such as base indices, restricted indices, and multiplicative weight-free indices.

The purpose of this chapter is to review recent literature on selection indices, and, by so doing, to provide a background on the types of selection problems for which selection indices have been considered. This review also serves to illustrate some of the methods used for estimating genotypic and phenotypic variances and covariances.

II. APPLICATIONS INVOLVING IMPROVEMENT OF A SINGLE QUANTITATIVE TRAIT

Smith[1] introduced the concept of a linear selection index by developing an index for improved grain yield of wheat, *Triticum aestivum*. He concluded that using an index based on yield, yield components, and straw weight would result in greater response in yield than would selection for yield itself. Similar conclusions have been reached by Murthy and Rao[2] in a study of lodging resistance in barley, *Hordeum vulgare*. They concluded that an index based on plant height and weight of the main shoot would give only 75% as much improvement in lodging angle as would selection for lodging angle itself. However, they considered that it is easier to measure plant height and main stem weight than lodging angle, and that the index would therefore provide a useful basis for improving lodging resistance.

Rosielle and Brown[3] assessed the use of selection indices for improving traits related to *Septoria nodorum* resistance in wheat. They evaluated yield, yield components, and disease symptoms on infected and disease-free (fungicide control) plots of F_2-derived F_3 lines from six wheat crosses. They found that using severity of disease on the flag leaf and on the head as secondary traits failed to provide much advantage in selecting for improved yield or increased seed weight under diseased conditions.

Thurling[4] evaluated several selection indices for improving yield of rapeseed, *Brassica campestris*, in western Australia. Two full-sib families within each of 32 half-sib families were evaluated in two blocks at a single location. Smith's[1] method was used to calculate several selection indices for improving yield and expected responses were compared. Indices based on yield components alone would not result in as much response as selection for yield *per se*. However, indices based on yield components and certain vegetative characteristics resulted in a higher expected response than selection for yield alone. The difficulty of conducting an extensive evaluation of a selection index over several environments led the author to question the practical value of selection indices.

Richards and Thurling[5] studied 112 full-sib families from bulk populations of two rapeseed species, *B. campestris* and *B. napus*. They grew the material under rainfed and irrigated conditions and compared predicted response to selection based on yield alone with predicted responses to selection based on several indices. In both species, expected yield increases were greater for indices which included yield and one or more correlated traits than for selection based on yield alone.

Paul et al.[6] calculated all possible selection indices involving yield and five other characteristics in brown mustard, *Brassica juncea*. Expected response in yield was greater for the selection indices only when yield was included in the index. It is disconcerting to note that expected responses from some of the selection indices which included yield were less than expected for selection based on yield alone. One must suspect that there was an error in calculation or in printing.

Selection indices for improved yield of soybean, *Glycine max*, were evaluated by Singh and Dalal.[7] F_3 and F_4-bulks of a diallel set of crosses among six soybean cultivars were tested in a randomized complete block experiment with four replications. Measurements on six characteristics, including yield, were made on 20 plants within each plot. Expected responses for all possible selection indices for improved seed yield were compared with expected response to direct selection for seed yield alone. None of the indices had an expected response exceeding that for yield alone by more than 6.4%. Of those indices which did not include yield, several gave expected responses which were nearly as high as that expected from direct selection for yield. The authors did not discuss the implications of estimating genotypic and phenotypic parameters from a mixed population of six inbred parents, 15 F_3 bulks and 15 F_4 bulks.

Pritchard et al.[8] also evaluated various selection indices for improving yield of soybean. Yield and yield component data, expressed on a logarithmic scale, were collected on 10 plants in each of five replications. Three selection indices were evaluated using F_3, F_4 and F_5 selections of two crosses in successive years. None of the three selection indices included yield itself. However, one would expect each to result in responses equal to selection for yield if the index coefficients had been set equal to unity. The authors[8] concluded that selection indices would give expected yield responses only marginally higher than direct selection for yield in the F_4. However, different results were obtained for selection in the F_5. The authors attributed this discrepancy to low heritability for yield in the F_5 and noted that actual response to direct selection for yield was similar in both generations.

Abo El-Zahab and El-Kilany[9] compared several selection strategies for increasing lint yield in cotton, *Gossypium barbadense*. They selected among F_2 and F_3 plants in one cross and evaluated F_3 and F_4 progeny rows in a replicated experiment in order to assess realized response to selection. Seven unrestricted indices, involving lint yield and its components, were evaluated. In general, realized response to selection was not as great as predicted. Of the unrestricted indices, one which included lint yield, lint/seed, and bolls/plant was expected to be most efficient for selection in F_2 and F_3. However, realized response for this index was greatest only in the F_3.

Several studies have been conducted to evaluate the efficiency of index selection in specialty crops.[10-12] In lentil, *Lens culinaris*, selection for yield and yield components was expected to be more efficient than selection for yield alone in a sample of 49 pure strains.[10] The same was true for cowpea,[11] *Vigna unguiculata*, and castorbean,[12] *Ricinus communis*.

In an evaluation of 67 potato, *Solanum tuberosum*, cultivars and hybrids, Gaur et al.[13] found that a selection index based on tuber yield, number of tubers per plant, and average tuber weight was expected to give a 62% greater response in tuber yield than direct selection for tuber yield alone. Tuber yield was positively correlated with average tuber weight. There was a strong negative correlation between average tuber weight and number of tubers per plant.

Singh and Baghel[14] evaluated several selection indices on 14 lines of sorghum, *Sorghum bicolor*. Of 12 indices involving yield and/or some of its five components, only those which included yield itself had expected responses equal to or greater than expected from direct selection for yield. None gave an expected response substantially greater than that for direct selection for yield.

Yousaf[15] gave a detailed account of the calculation of selection indices for improving grain yield in a population of maize, *Zea mays*. The experiment included 10 groups of 25 F_1 hybrids among S_1 lines of one synthetic cultivar. Each group consisted of hybrids among five lines used as males and another five used as females. A North Carolina Design II analysis was used to estimate additive genetic and phenotypic variances and covariances. For improving grain yield, expected responses were greatest for those indices which included grain yield in the index. An index which included number of rows per ear, number of kernels per row, and kernel weight was expected to be nearly as efficient as the best index and about 10% more efficient than selection for yield alone.

Subandi et al.[16] conducted a similar evaluation in two maize crosses. Indices, based on grain yield, percent stalk lodging, and percent dropped ears, gave essentially no expected improvement over selection for yield alone. All three characters are important in the determination of harvestable yield. The authors calculated the expected response in harvestable yield and showed that all indices were expected to be more efficient than direct selection for any one component.

Subandi et al.[16] also showed that a multiplicative index, based on the methodology of Elston,[17] was expected to be nearly as efficient as one based on desired gains,[18] or one where relative economic values were set at 1.0 for grain yield and 0.0 for the other two traits. Because of the simplicity of the multiplicative index, they recommended it for improving harvestable yield in maize. In an earlier study, Robinson et al.[19] showed that several indices involving yield, yield components, and/or plant height resulted in greater expected response than direct selection for yield in maize.

Widstrom[20] developed selection indices for resistance to corn earworm in maize. The indices were applied to four cycles of recurrent selection.[21] The author concluded that none of the indices was superior to conventional selection for earworm resistance. The author did suggest, however, that a selection index might be useful when considering all the characteristics that are important in a breeding program.

Singh[22] calculated several selection indices for seed yield in alfalfa, *Medicago sativa*, and concluded that none would be more efficient than direct selection for seed yield. These results were based on evaluation of progeny of a diallel cross among seven selected lines from a cross between two alfalfa cultivars.

Moreno-Gonzalez and Hallauer[23] developed a theory for combining data from full-sib families and related S_2 families in each of two parental populations in order to increase the response to full-sib reciprocal recurrent selection in maize. For traits with low heritability ($h^2 < 0.5$), combined reciprocal recurrent selection will be more efficient than unmodified reciprocal recurrent selection, even when differences in amount of effort are taken into consideration.

England[24] used a selection index approach to provide guidelines for selecting within inbred species such as barley, *Hordeum vulgare*. Based on relative sizes of within- and among-family variances, England demonstrated that the number of plants per plot should be kept as small as permitted by agronomic considerations. His work demonstrated that selection indices could be used to develop optimum weighting of family means and within-family deviations in inbred species.

Walyaro and Van der Vossen[25] used selection indices to select for average yield of coffee, *Coffea arabica*, assessed over a number of years. In coffee, yield is usually recorded as average yield per plant over a number of years. In this study, estimated response to selection for average yield over 10 years was compared to response expected from indices based on yield, yield components, and various other measurements taken in the first two years of the nursery. The nursery consisted of 16 cultivars, each planted in five replications with ten trees per plot. The authors estimated that selection based on an index of characteristics measured in the first two years would be 97% as efficient as direct selection for 10-year average yield.

Namkoong and Matzinger[26] tried to apply the theory of selection indices to the improvement of the annual growth curves of *Nicotiana tabacum*. A growth curve can be considered as a series of highly correlated traits all of which require improvement. Namkoong and Matzinger used plant height at eight weekly intervals to estimate the growth curve. A diallel cross among the eight parents of the base population provided estimates of the genotypic variances and covariances among the eight height measurements. Relative economic weights were set equal to the mean height at each of the eight assessment times, and a selection index was calculated. The estimated index contained both positive and negative coefficients and, when applied over four cycles of selection, gave an increase in final height of only 1.9%. An alternative approach, that of a selection index based on estimated coefficients of a three-parameter growth curve, appeared more efficient.

In an evaluation of selection indices for sugarcane, *Saccharum* spp., Miller et al.[27] studied clones from 50 F_1 seedlings in each of four biparental crosses. Clones were planted in two replications at each of two locations in each of two years. When stands were 9 to 14 months of age, they were evaluated for six characteristics. Two sets of indices, one for improved cane yield and one for improved sucrose yield, were investigated. The primary trait was not included in the index in either case. Selection indices provided expected responses of 89 and 92% of that expected for direct selection for cane yield and sucrose yield, respectively.

III. APPLICATIONS INVOLVING THE SIMULTANEOUS IMPROVEMENT OF TWO OR MORE TRAITS

Eagles and Frey[28] compared several methods of selecting for simultaneous improvement of grain yield and straw yield of oat, *Avena sativa*. They argued that both grain and straw are economically important, and considered that the relative economic value for grain should be twice that for straw. In their experiment, 1200 F_9-derived lines from a composite population were evaluated in replicated tests at three sites over two years. Data from the second year were used to test the validity of expectations based on first year data. Due to the effects of genotype-environment interaction on single year estimates, observed responses were less than expected in all cases. However, both expected and actual response ranked selection methods in the same general order. Economic values (2 × grain yield + straw yield) usually increased most by use of an index or with independent culling at carefully chosen culling levels. Two indices, the optimum index of Smith[1] and the base index of Williams,[29] gave similar results. Since the base index requires only that relative economic weights be used as index coefficients, the authors preferred this simpler index. This conclusion was similar to that of Elgin et al.[30] who looked at simultaneous improvement of five traits in alfalfa.

Rosielle and Frey[31] also worked with F_9-derived oat lines and evaluated the use of restricted selection indices[32] for improving grain yield while restricting the correlated responses in heading date and plant height. Because all three traits are positively correlated, any restrictions involving zero or negative increases in heading date or plant height reduced expected response in yield. The authors suggested that this effect could be reduced by including harvest index (grain to straw ratio) as a secondary trait in the selection index. Rosielle et al.[33] obtained similar results when applying restricted indices to selection for economic value.

Suwantaradon et al.[34] evaluated several types of selection indices for simultaneous improvement of seven characteristics in S_1 maize lines. Besides the optimum index selection method,[1] they used a base index[29] and the modified index of Pesek and Baker.[18] The optimum index gave expected responses for individual traits that were not closely related to the assigned economic values. The same was true for the base index. In fact, the base index and the optimum index gave similar results and prompted the authors to conclude that, since it is much easier to use, the base index would be preferred if exact relative economic values were known. The modified index[18] was expected to give individual trait responses proportional to desired gains. However, the authors calculated that desired gains could be achieved only after 14 cycles ($= 28$ years) of selection. Nevertheless, the authors recommended the modified index when exact weights are not known.

A similar study of simultaneous improvement of yield and protein concentration in maize was conducted by Kauffmann and Dudley.[35] These authors concluded that the modified "desired gains" index[18] was feasible for simultaneous improvement of yield and protein concentration in maize.

Subandi et al.[16] compared several methods of index selection in maize and proposed the use of a multiplicative index that estimates dry grain yield on standing plants. Compton and Lonnquist[36] showed that the proposed index was effective over four cycles of full-sib recurrent selection.

Crosbie et al.[37] compared several types of selection indices for simultaneous improvement of cold-tolerance traits in maize. They indicated that the modified index of Pesek and Baker[18] was not satisfactory because expected performance was particularly poor when large gains were specified for one trait and small gains for the others. A similar criticism was levied against the base and optimum indices where individual trait responses did not seem to be concordant with desired responses. The authors concluded that a better method might be the mutiplicative index of Elston,[17] or its linear approximation.[38] They also recommended the use of a rank summation method for evaluating several traits at a time.

Motto and Perenzin[39] evaluated selection indices for the simultaneous improvement of yield and protein concentration in a synthetic population of maize. They tested 163 S_1 lines in three replications at two locations, and concluded that the modified index of Pesek and Baker[18] would be useful for selecting for these two traits.

St. Martin et al.[40] calculated restricted selection indices for increasing grain yield in maize while maintaining kernel hardness, moisture concentration, protein concentration, and protein quality at satisfactory levels. An experiment consisting of 112 full-sib families in seven sets of 16 entries (four males × four females) was used to estimate genotypic and phenotypic parameters. The authors concluded that the use of selection indices for simultaneous improvement of several traits should probably be accompanied by a certain amount of subjective judgement by the breeder. Comparisons of expected responses are not very reliable because of the large sampling errors associated with such estimates.

Radwan and Momtaz[41] evaluated several optimum indices for the simultaneous improvement of seed yield, straw yield, and earliness in flax, *Linum usitatissimum*. Estimates of genotypic and phenotypic variances and covariances were obtained from data on the two parents and on the F_2, F_3, F_4 and F_5. Realized responses showed that a selection index combining all three traits was more efficient than single trait selection for improving overall value.

Sullivan and Bliss[42] used the modified "desired gains" index[18] to improve seed yield and protein concentration in a recurrent mass selection experiment with common bean, *Phaseolus vulgaris*. These authors used the data from parental lines to estimate environmental variances and covariances. They suggested modifications to their methodology which would result in better estimates of the index coefficients.

Davis and Evans[43] evaluated 15 characteristics of 112 F_4 lines of navy bean, *Phaseolus vulgaris*, from different crosses in a completely randomized design with one to three replications of each line. Genotypic and phenotypic variances and covariances were estimated from the among-line and within-line mean squares and cross-products. An optimum selection index[1] was calculated in which relative economic weights were 1.0 for seed yield, -0.5 for seed shape, -0.5 for time to maturity, 1.0 for height of pod tip, and 0.0 for the remaining 11 characteristics. They estimated that a base index,[29] where economic weights were used as index coefficients, would be 73% as efficient as an optimum index based only on the primary traits. More complex indices, using some of the secondary traits, were expected to be even more efficient. They concluded that data on certain plant type characteristics would be more useful than data on yield components in developing indices for improving yield.

IV. SUMMARY

This review of recent literature on selection indices serves as a guide to current thinking about the value of the technique, as well as indicating the breadth of the methodology being considered. Much of the literature is oriented toward maximization of yield. Often indices developed for this purpose are based on yield components as well as yield itself. There is a lack of clarity about what can be expected if both yield and its components are included in a selection index for improving yield itself. In the first place, expected response to index selection for yield has to be greater than or equal to the expected response to direct selection for yield, if yield and one or more of its components are included in the index and if the same selection intensity is assumed under both selection strategies.

A second consideration in using both yield and its components in an index is the fact that yield is equal to the product of its components. If the data on yield and its components are expressed on a logarithmic basis, the transformed yield data should equal the sum of the transformed data for its components. Inclusion in an index of transformed yield data and the transformed data for all its components is repetitious. In fact, inclusion of yield and all of its components should theoretically make it impossible to develop a solution to the resulting index equations. In practice, this equality fails to hold because of measurement errors in yield and its components. For this reason, the relationship is often masked. Including one or more components of yield, but not all, may lead to enhanced response in yield by permitting greater or lesser weight to be applied to one or more components. Theoretically, the most efficient index for improving a complex trait is one consisting of all of its components but excluding the complex trait itself.

The problem of assigning relative economic weights arises in most attempts to develop indices for the simultaneous improvement of two or more traits. In some instances, researchers have assigned weights in a rather arbitrary manner and have been discouraged by the expected response to the resulting index selection. Where each of the primary traits can be related directly to the economic value of the crop, the task of assigning weights is not difficult. In these cases, selection indices are expected to enhance overall response to selection.

Several authors have considered the modified "desired gains" index of Pesek and Baker[18] (shown in Chapter 1 to be a special case of the restricted index of Tallis[32]). In one case, the authors calculated that the specified objectives would be reached only after an unacceptable number of cycles of selection. The exercise of specifying desired gains may be useful in making the breeder aware of whether or not the expectations for improvement are in line with the genetical limitations of the population undergoing selection.

Several authors have shown that a base index, where the relative economic weights are used as index coefficients, often gives expected responses that are similar to those of the optimum selection index. If this were always the case, a base index would be preferred because its use does not require estimates of genotypic and phenotypic variances and covariances. However, a base index will not be able to include secondary traits which might further enhance response. This is not a great weakness if secondary traits show little correlation with primary traits. If the index includes only primary traits, the base index is expected to give results similar to the optimum index, provided the genotypic variances and covariances are a constant (or nearly constant) multiple of the corresponding phenotypic variances and covariances. In cases where estimates of genotypic variances and covariances are unreliable, it may be better to proceed with that assumption and use a base index.

Recent literature also shows a variety of methods for estimating genotypic and phenotypic variances and covariances. Samples of inbred or partially inbred lines, normally assumed to be random, can be tested in replicated experiments. In some cases, replication will also include replication over sites and over seasons. Methods for extracting estimates of genotypic variances and covariances from the analysis of such data are well documented. However, there are no clear guidelines on sample sizes required for reliable estimation. Sample size and extent of replication vary considerably from experiment to experiment.

In some cases, researchers have used data from homogeneous and segregating populations in order to separate the effects of genotype and environment on the variation and covariation between traits. This approach is less popular than the evaluation of partially inbred lines in replicated experiments.

In maize, several very thorough studies have been based on the North Carolina Design II method. This method allows the identification of additive genotypic causes of variation and covariation. These studies have included several sets of progeny developed by making all possible crosses between a number of female parents and a number of male parents. This method is preferred for estimating variances and covariances in crosspollinated crops.

In reviewing some of the recent literature, one finds little evidence that selection indices are being used routinely for crop improvement. These techniques apparently are not an important part of present day crop improvement programs. Nevertheless, published literature tends to demonstrate that selection could be made more effective by adopting some objective method of weighting traits when several are to be improved simultaneously. Modern computing power can eliminate much of the drudgery of calculating and applying selection indices in plant breeding programs. However, if index methods require that large volumes of data be collected to develop a useful index, the benefits may not justify the additional effort.

REFERENCES

1. **Smith, H. F.,** A discriminant function for plant selection, *Ann. Eugenics,* 7, 240, 1936.
2. **Murthy, B. N. and Rao, M. V.,** Evolving suitable index for lodging resistance in barley, *Indian J. Genet. Plant Breeding,* 40, 253, 1980.
3. **Rosielle, A. A. and Brown, A. G. P.,** Selection for resistance to *Septoria nodorum* in wheat, *Euphytica,* 29, 337, 1980.
4. **Thurling, N.,** An evaluation of an index method of selection for high yield in turnip rape, *Brassica campestris* L. ssp. *oleifera* Metzg., *Euphytica,* 23, 321, 1974.
5. **Richards, R. A. and Thurling, N.,** Genetic analysis of drought stress response in rapeseed *(Brassica campestris* and *B. napus).* II. Yield improvement and the application of selection indices, *Euphytica,* 28, 169, 1979.
6. **Paul, N. K., Joarder, O. I., and Eunus, A. M.,** Genotypic and phenotypic variability and correlation studies in *Brassica juncea* L., *Z. Pflanzenzuecht.,* 77, 145, 1976.
7. **Singh, C. B. and Dalal, M. A.,** Index selection in soybean, *Indian J. Genet. Plant Breeding,* 39, 234, 1979.
8. **Pritchard, A. J., Byth, D. E., and Bray, R. A.,** Genetic variability and the application of selection indices for yield improvement in two soya bean populations, *Aust. J. Agric. Res.,* 24, 81, 1973.
9. **Abo El-Zahab, A. A. and El-Kilany, M. A.,** Evaluation of several selection procedures for increased lint yield in segregating generations of Eygptian cotton *(Gossypium barbadense* L.), *Beitr. Trop. Landwirtsch. Veterinaermed.,* 17, 153, 1979.

10. **Nandan, R. and Pandya, B. P.**, Correlation, path coefficient and selection indices in lentil, *Indian J. Genet. Plant Breeding*, 40, 399, 1980.

11. **Tikka, S. B. S. and Asawa, B. M.**, Selection indices in cowpea, *Vigna unguiculata* (L.) Walp., *Pulse Crops Newsletter*, 1, 68, 1981.

12. **Sulih, S. H. and Khidir, M. O.**, Correlations, path analyses and selection indices for castorbean *(Ricinus communis* L.), *Expl. Agric.*, 11, 145, 1975.

13. **Gaur, P. C., Kishore, H., and Gupta, P. K.**, Studies on character association in potatoes, *J. Agric. Sci.*, 90, 215, 1978.

14. **Singh, R. P. and Baghel, S. S.**, Yield components and their implication to selection in sorghum, *Indian J. Genet. Plant Breeding*, 37, 62, 1977.

15. **Yousaf, M.**, The use of selection indices in maize (*Zea mays* L.), in *Genetic Diversity in Plants*, Muhammed, A., Askel, R., and von Borstel, R. C., Eds., Plenum Press, New York, 1977, 259.

16. **Subandi, Compton, W. A., and Empig, L. T.**, Comparison of the efficiencies of selection indices for three traits in two variety crosses of corn, *Crop Sci.*, 13, 184, 1973.

17. **Elston, R. C.**, A weight-free index for the purpose of ranking or selection with respect to several traits at a time, *Biometrics*, 19, 85, 1963.

18. **Pesek, J. and Baker, R. J.**, Desired improvement in relation to selection indices, *Can. J. Plant Sci.*, 49, 803, 1969.

19. **Robinson, H. F., Comstock, R. E., and Harvey, P. H.**, Genotypic and phenotypic correlations in corn and their implications in selection, *Agron. J.*, 43, 282, 1951.

20. **Widstrom, N. W.**, Selection indexes for resistance to corn earworm based on realized gains in corn, *Crop Sci.*, 14, 673, 1974.

21. **Widstrom, N. W., Wiseman, B. R., and McMillan, W. W.**, Responses to index selection in maize for resistance to ear damage by the corn earworm, *Crop Sci.*, 22, 843, 1982.

22. **Singh, S. M.**, Genetic basis of seed setting in alfalfa, *Theor. Appl. Genet.*, 51, 297, 1978.

23. **Moreno-Gonzalez, J. and Hallauer, A. R.**, Combined S$_2$ and crossbred family selection in full-sib reciprocal recurrent selection, *Theor. Appl. Genet.*, 61, 353, 1982.

24. **England, F.**, Response to family selection based on replicated trials, *J. Agric. Sci.*, 88, 127, 1977.

25. **Walyaro, D. J., and Van der Vossen, H. A. M.**, Early determination of yield potential in arabica coffee by applying index selection, *Euphytica*, 28, 465, 1979.

26. **Namkoong, G. and Matzinger, D. F.**, Selection for annual growth curves in *Nicotiana tabacum* L., *Genetics*, 81, 377, 1975.

27. **Miller, J. D., James, N. I., and Lyrene, P. M.**, Selection indices in sugarcane, *Crop Sci.*, 18, 369, 1978.

28. **Eagles, H. A. and Frey, K. J.**, Expected and actual gains in economic value of oat lines from five selection methods, *Crop Sci.*, 14, 861, 1974.

29. **Williams, J. S.**, The evaluation of a selection index, *Biometrics*, 18, 375, 1962.

30. **Elgin, J. H., Jr., Hill, R. R., Jr., and Zeiders, K. E.**, Comparison of four methods of multiple trait selection for five traits in alfalfa, *Crop Sci.*, 10, 190, 1970.

31. **Rosielle, A. A. and Frey, K. J.**, Application of restricted selection indices for grain yield improvement in oats, *Crop Sci.*, 15, 544, 1975.

32. **Tallis, G. M.**, A selection index for optimum genotype, *Biometrics*, 18, 120, 1962.

33. **Rosielle, A. A., Eagles, H. A., and Frey, K. J.**, Application of restricted selection indexes for improvement of economic value in oats, *Crop Sci.*, 17, 359, 1977.

34. **Suwantaradon, K., Eberhart, S. A., Mock, J. J., Owens, J. C., and Guthrie, W. D.**, Index selection for several agronomic traits in the BSSS2 maize population, *Crop Sci.*, 15, 827, 1975.

35. **Kauffmann, K. D. and Dudley, J. W.**, Selection indices for corn grain yield, percent protein, and kernel weight, *Crop Sci.*, 19, 583, 1979.

36. **Compton, W. A. and Lonnquist, J. H.**, A multiplicative selection index applied to four cycles of full-sib recurrent selection in maize, *Crop Sci.*, 22, 981, 1982.

37. **Crosbie, T. M., Mock, J. J., and Smith, O. S.**, Comparison of gains predicted by several selection methods for cold tolerance traits in two maize populations, *Crop Sci.*, 20, 649, 1980.

38. **Baker, R. J.**, Selection indexes without economic weights for animal breeding, *Can. J. Anim. Sci.*, 54, 1, 1974.

39. **Motto, M. and Perenzin, M.**, Index selection for grain yield and protein improvement in an opaque-2 synthetic maize population, *Z. Pflanzenzuecht.*, 89, 47, 1982.

40. **St. Martin, S. K., Loesch, P. J., Jr., Demopulos-Rodriguez, J. T., and Wiser, W. J.**, Selection indices for the improvement of opaque-2 maize, *Crop Sci.*, 22, 478, 1982.

41. **Radwan, S. R. H. and Momtaz, A.**, Evaluation of seven selection indices in flax *(Linum usitatissimum* L.), *Egypt. J. Genet., Cytol.*, 4, 153, 1975.

42. **Sullivan, J. G. and Bliss, F. A.,** Recurrent mass selection for increased seed yield and seed protein percentage in common bean *(Phaseolus vulgaris* L.) using a selection index, *J. Am. Soc. Hortic. Sci.,* 108, 42, 1983.
43. **Davis, J. H. C. and Evans, A. M.,** Selection indices using plant type characteristics in navy beans *(Phaseolus vulgaris* L.), *J. Agric. Sci.,* 89, 341, 1977.

Chapter 3

BASIC CONCEPTS IN QUANTITATIVE GENETICS

I. INTRODUCTION

An understanding of how selection indices can be developed, and how they can aid in crop improvement, will be easier if one first understands some of the quantitative genetic principles that are germain to the subject. This chapter is designed to review the quantitative genetic principles of three topics which are critical to the application of selection indices and which are also central to the overall topic of artificial selection. These topics include (1) methods used for estimating unobservable genotypic values, (2) the general principles of heritability, coheritability, and response to selection, and (3) methods used for estimating genotypic covariances and correlations.

The approach in this chapter is to discuss all three topics from a uniform perspective — that of linear regression. For this reason, some of the presentation may seem unfamiliar. Hopefully, the general approach will provide a basis for developing an understanding of more complex matters to be covered in succeeding chapters.

In this and following chapters, it will be useful to understand the concepts of variances and covariances of linear functions of variables. Let $Z = aX + bY$ be a linear function of the variables X and Y and the known constants a and b. The variance of the linear function, Z, can be expressed in terms of the variances and covariances of its components. Thus,

$$\sigma_Z^2 = a^2\sigma_X^2 + 2ab\sigma_{XY} + b^2\sigma_Y^2$$

If X and Y are independent (uncorrelated), $\sigma_{XY} = 0$ and

$$\sigma_Z^2 = a^2\sigma_X^2 + b^2\sigma_Y^2$$

For example, if the phenotypic value, P, is a linear function of a genotypic value, G, and an environmental deviation, E; and if G and E are uncorrelated, the phenotypic variance is given by

$$\sigma_P^2 = \sigma_G^2 + \sigma_E^2$$

If, in addition to $Z = aX + bY$, there is a second linear function, $W = cX + dY$, the covariance between the two linear functions is given by

$$\sigma_{WZ} = ac\sigma_X^2 + (bc + ad)\sigma_{XY} + bd\sigma_Y^2$$

With X and Y uncorrelated,

$$\sigma_{WZ} = ac\sigma_X^2 + bd\sigma_Y^2$$

Then, if one considers that $P = G + E$ and $G = G$ are two linear functions, the method indicated above shows that $\sigma_{GP} = \sigma_G^2 + \sigma_{GE} = \sigma_G^2$

II. ESTIMATION OF GENOTYPIC VALUES

Genotypic value is the average phenotypic performance of a particular genotype when it can be evaluated in a complete reference set of environments.[1] Thus, for a particular genotype,

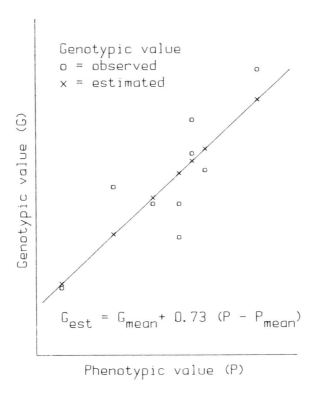

FIGURE 1.   The relationship between hypothetical genotypic and
phenotypic values showing estimated genotypic values based on
the linear regression on phenotypic values (b_{GP} = heritability =
0.73).

its genotypic value might be considered as the hypothetical average performance over several years within a particular geographic region.

In most cases, genotypic values cannot be observed directly. The basic concept used to estimate the genotypic value of a quantitative trait is to recognize that the observed phenotypic performance of a particular individual or line is made up of two components, a component (G) related to the genotypic constitution of that individual or line, and a component (E) related to the particular environmental situation under which it is observed. These two components are assumed to act additively in their effect on the observed phenotypic response, i.e.,

$$P_i = G_i + E_i$$

where P_i, G_i, and E_i represent the phenotypic value, the genotypic value, and the environmental deviation associated with the ith individual or line. For the two components to act in an additive manner, the underlying biology of the trait must be such that the effects are additive, and the genotypes must be properly randomized with respect to environmental effects. The latter requirement can usually be met in plant populations. The former will be discussed in Chapter 5.

Two basic methods are available for estimating the value of a particular genotype. In the first place, the "best" estimate of the genotypic value of an individual is considered to be that value which will show, on average, the least deviation from the true value. The best estimates of genotypic values are those estimated by the linear regression of genotypic values on phenotypic values (Figure 1).

By definition, the linear regression of genotypic value on phenotypic value is

$$b_{GP} = \frac{\sigma_{GP}}{\sigma_P^2}$$

where σ_{GP} is the covariance between genotypic and phenotypic values, and σ_P^2 is the phenotypic variance. Since $P_i = G_i + E_i$,

$$\sigma_{GP} = \sigma_{(G)(G+E)} = \sigma_G^2 + \sigma_{GE}$$

Furthermore, if G_i and E_i are independent, $\sigma_{GE} = 0$, $\sigma_{GP} = \sigma_G^2$ and

$$b_{GP} = \frac{\sigma_G^2}{\sigma_P^2}$$

The latter expression is known as heritability. Using the principle of linear regression, one would estimate the genotypic value, G_i, of the ith individual or line having phenotypic value P_i as

$$G_i = \overline{G} + b_{GP}(P_i - \overline{P})$$

By convention, the E_is are considered to sum to zero and $\overline{G}$ is equal to $\overline{P}$. Thus,

$$G_i = \overline{P} + b_{GP}(P_i - \overline{P})$$

An estimate of the regression of genotypic value on phenotypic value, b_{GP}, requires an estimate of the variance of genotypic values as well as an estimate of the variance of phenotypic values. The latter is available directly from measurement of the phenotypic values of a sample of individuals from the reference population. Genotypic variance can be estimated by recognizing that, if genotypic effects and environmental deviations are uncorrelated, the phenotypic variance will be the sum of the genotypic variance and the environmental variance. Thus, if one could estimate the phenotypic variance and the environmental variance, the genotypic variance could be estimated by the difference between the phenotypic and environmental variances. To estimate the environmental variance requires repeated measurements on the same genotype tested over a sample of environments. If inbred lines are being studied, an estimate of the environmental variance will be available directly from an analysis of the repeated measures on each inbred genotype. With segregating material, it will be necessary to include genotypically homogeneous material (inbred parents, F_1s, or clones) to afford estimates of environmental variation.

Given estimates of the phenotypic (σ_P^2) and environmental (σ_E^2) variances, the genotypic variance is estimated as $\sigma_G^2 = \sigma_P^2 - \sigma_E^2$ and the regression of genotypic value on phenotypic value can be estimated as

$$b_{GP} = \frac{\sigma_P^2 - \sigma_E^2}{\sigma_P^2}$$

Then, the genotypic value of the ith genotype can be estimated as

$$G_i = \overline{P} + b_{GP}(P_i - \overline{P}) \tag{2}$$

There is a very real tendency to think of phenotypic values as being dependent upon the underlying genotypic values rather than vice versa as indicated above and in Figure 1. However, in relation to crop improvement, the change in unobservable genotypic values will be very much dependent upon their relationship with observable phenotypic values. Since selection pressure will be exerted through selection based on observed phenotypic values, any change in genotypic value will indeed be dependent upon change in phenotypic value. In this sense, it is reasonable to regard genotypic values as dependent upon phenotypic values.

Under the assumptions stated above, the regression coefficient cannot exceed 1.0 (when $\sigma_E^2 = 0$) nor be less than 0.0 (when $\sigma_G^2 = 0$). The regression line (Figure 1) will pass through the mean genotypic value and the mean phenotypic value by virtue of the assumption that the environmental deviations have an average value of zero. Because of this, and because the regression of genotypic value on phenotypic value must lie between 0 and 1, estimated genotypic values of particular individuals will be obtained by adjusting high phenotypic values downward and low phenotypic values upward.

This statement can be illustrated by rewriting the prediction equation (Equation 2) as follows:

$$(G_i - \overline{G}) = b_{GP}(P_i - \overline{P}) = (P_i - \overline{P}) + (b_{GP} - 1)(P_i - \overline{P})$$

Then, since $\overline{G} = \overline{P}$,

$$G_i = P_i + (b_{GP} - 1)(P_i - \overline{P})$$

To estimate G_i whenever P_i exceeds $\overline{P}$, P_i will be reduced by an amount $(P_i - \overline{P})$ $(b_{GP} - 1)$ unless $b_{GP} = 1$ (i.e., heritability = 100%). If, on the other hand, P_i is less than $\overline{P}$, P_i will be increased by an amount equal to $(P_i - \overline{P})$ $(b_{GP} - 1)$ to estimate G_i. Thus, if heritability is less than 100%, the ''best'' estimate of the genotypic value of a particular individual will be obtained by adjusting its phenotypic value toward the population mean by an amount which will depend upon heritability and how different the individual is from the mean.

A second method for estimating genotypic value is through the use of progeny testing. This is the only method available to animal breeders and is often the preferred method when working with cross-pollinated species which are propagated primarily by seed. If parents are randomly mated, the average phenotypic value of progeny of a particular parent will deviate from the population mean by one-half the amount by which the genotypic value of the parent deviates from the population mean. In this context, genotypic value refers to the additive genetic value (or breeding value), a property of the alleles passed from parent to progeny, in contrast to total genotypic value which reflects all of the effects attributed to the genotype of the individual.

The use of parent-offspring relationships to estimate genotypic values of the parents is similar in concept to the method discussed above. With reference to Figure 1, G_i can refer to the genotypic value of the parent while P_i refers to the phenotypic value of one of its progeny. Using the same methods of linear regression, the estimate of the genotypic value of the parent is given by

$$G_i = \overline{G} + b_{GP}(P_i - \overline{P})$$

In this case, the phenotypic value of the progeny is made up of a genotypic value equal to one-half the genotypic value of its mother, plus a genotypic value equal to one-half the genotypic value of its father, plus an environmental deviation, i.e.,

$$P_i = 0.5\ G_i + 0.5\ G_j + E_i$$

With random mating, G_i and G_j will be uncorrelated so that the covariance between genotypic value of the parent and the phenotypic value of its progeny will be

$$\sigma_{GP} = \sigma_{G_i(0.5\ G_i + 0.5\ G_j + E_i)} = 0.5\ \sigma_G^2 \text{ and}$$

$$b_{GP} = \frac{\sigma_{GP}}{\sigma_P^2} = 0.5\ \frac{\sigma_G^2}{\sigma_P^2}$$

The estimated genotypic value of the parent is thus given by

$$G_i = \overline{G} + 0.5\ \frac{\sigma_G^2}{\sigma_P^2}\ (P_i - \overline{P})$$

where σ_G^2 refers to the additive genetic variance in the parent or offspring generation.

This requires some discussion in light of the usual definition of breeding value of an individual. The breeding value of an individual is defined as twice the difference between the mean of its half-sib progeny and the mean of the total offspring population. If so, the genotypic value of the parent should be predicted by multiplying the deviation between the phenotypic value of its progeny and the population mean by 2.0 rather than 0.5 h^2 (which can never exceed 0.5). To resolve this apparent discrepancy requires careful consideration of the phenotypic variances among individual progeny and among means of large half-sib families. The phenotypic variance among individual progeny will equal the phenotypic variance in the parent population. Since $\sigma_{GP} = 0.5\ \sigma_G^2$ (see above), the regression of parent breeding value on individual progeny phenotypic values is $b_{GP} = 0.5\ \sigma_G^2/\sigma_P^2$ which cannot exceed 0.5. If, however, one considers $P_i.$ to be the average phenotypic value of all possible progeny of the ith parent, the expected value of $P_i.$ will be 0.5 G_i since the genetic effects contributed by the other parents are assumed to sum to zero, as are the environmental deviations. In this case, the variance of the mean phenotypic values will be $\sigma_P^2. = 0.25\ \sigma_G^2$ and the regression of parent genotypic values on these means will be

$$b_{GP} = \frac{0.5\sigma_G^2}{0.25\sigma_G^2} = 2.0$$

Thus, by the usual definition of breeding value, the breeding value of a parent, expressed as a deviation from the population mean $(G_i - G)$, can be estimated as twice the difference between its (infinitely large) half-sib progeny mean and the mean of the progeny population. However, if one can measure only one progeny from each parent, the best estimate of parental breeding value is given by multiplying the progeny deviation by $0.5b_{GP} = 0.5h^2$.

To estimate the regression of parent breeding value on progeny phenotypic value, one would normally calculate the regression of the offspring phenotypic value, O, on the parent phenotypic value, P. This regression, b_{OP}, is expected to equal 0.5 σ_G^2/σ_P^2 or one-half the heritability. Thus, heritability can be estimated as twice the regression of offspring on one parent. However, expected response to selection based on individual progeny performance is

$$(G_s - \overline{G}) = b_{OP}(P_s - \overline{P}) = 0.5\ h^2\ (P_s - \overline{P})$$

as implied above. If selection is based on the mean of half-sib or full-sib families, a similar approach can be used and appropriate modifications can be made to the formula for predicted response to selection.

Genotypic value cannot be measured directly. The methods of linear regression, with perhaps indirect estimates of appropriate covariances and variances, can be used to predict individual genotypic values as functions of observed phenotypic values. These predicted values are not very useful in practice. However, the approach does allow prediction of average response to selection and does form a central part of the theory relating to artificial selection in plant populations.

III. COMPARISON OF VARIOUS DEFINITIONS OF HERITABILITY

Although heritability is usually defined as the ratio of genotypic variance to phenotypic variance, it was shown in section II that heritability is also equal to the regression of genotypic value on phenotypic value. An alternative way to look at heritability is by considering the coefficient of determination of the regression of genotypic value on phenotypic value. If $P_i = G_i + E_i$ and $(G_i - \overline{G}) = b_{GP} (P_i - \overline{P})$, the coefficient of determination for the regression of genotypic value on phenotypic value is given by

$$r^2 = \frac{[\sigma_{GP}]^2}{\sigma_G^2 \sigma_P^2} = \frac{[\sigma_G^2]^2}{\sigma_G^2 \sigma_P^2} = \frac{\sigma_G^2}{\sigma_P^2} = h^2$$

From this formulation, one could define heritability as the proportion of the variation in genotypic values which can be explained by linear regression on phenotypic values. Since the coefficient of determination would be the same if phenotypic values were regressed on genotypic values, one might also define heritability as the proportion of variation in phenotypic values which can be explained by linear regression on genotypic values. While the latter definition is more common, the inability to measure actual genotypic values makes this definition of somewhat limited value.

Heritability can be estimated by the two basic methods outlined in Section II. The first method, based on estimates of variance components, usually gives estimates of the ratio of total genotypic variance to phenotypic variance; a ratio which is often referred to as "broad-sense" heritability. The second method is based on correlation between relatives and gives estimates of the ratio of additive genetic variance to phenotypic variance, a ratio that is usually referred to as "narrow-sense" heritability. The discussion in the previous section shows that broad-sense heritabilities are appropriate for dealing with problems concerning selection among inbred genotypes while narrow-sense heritabilities are more appropriate for problems concerning selection in random mating populations.

In Section II, it was suggested that genotypic values could be estimated by

$$G_i = \overline{G} + b_{GP} (P_i - \overline{P})$$

In a selection program, it is usual practice to select a number of genotypes which express phenotypic values higher than some chosen truncation point, and to be concerned about the average value of the selected genotypes. The expected average genotypic value of s selected genotypes would be equal to the average of the expected genotypic values for each individual, i.e.,

$$G_s = \frac{1}{s} \sum G_i = \overline{G} + \frac{1}{s} \sum b_{GP} (P_i - \overline{P}) = \overline{G} + b_{GP} \times \frac{1}{s} \sum (P_i - \overline{P})$$

$$= \overline{G} + b_{GP}(P_s - \overline{P}) \text{ and } \Delta G = (G_s - \overline{G}) = b_{GP} (P_s - \overline{P})$$

where G_s is the mean genotypic value of the s selected individuals and P_s is the mean phenotypic value of those selected individuals. $\overline{G}$ and $\overline{P}$ represent the average genotypic and phenotypic values of the unselected population. This equation is the familiar one that states that response to single trait selection is expected to equal the product of heritability (i.e., the linear regression of genotypic value on phenotypic value, b_{GP}) and the selection differential, $(P_s - \bar{P})$.

A third method of estimating heritability is by conducting a selection experiment and dividing the observed response by the selection differential. This method is really a special case of using correlations between relatives. The average values of the offspring of the selected parents are compared with the means of their parents in a special type of parent-offspring regression.

In this discussion, care has been taken to emphasize that the estimation of genotypic values need not be based on estimates of heritability. The use of the regression coefficient may seem awkward, but it provides a more general framework which can be extended to other problems of artificial selection. When considering several traits in a breeding program, it is possible to consider estimating genotypic values of one trait, say trait Y, from observations of phenotypic values on another trait, say trait X. By using the regression approach, the following prediction equation can be developed.

$$(G_{sY} - \bar{G}_Y) = b_{G(Y)P(X)} (P_{sX} - \bar{P}_X)$$

In this case, the regression coefficient is equal to

$$b_{G(Y)P(X)} = \frac{\sigma_{G(Y),[G(X)+E(X)]}}{\sigma^2_{P(X)}} = \frac{\sigma_{G(XY)}}{\sigma^2_{P(X)}}$$

the ratio of the genotypic covariance between the two traits to the phenotypic variance of trait X. This term is sometimes referred to as "coheritability".[2]

IV. ESTIMATION OF GENOTYPIC CORRELATIONS

Genotypic values for different traits may be correlated because they are influenced in part by genes which affect both traits or because they are influenced by different genes which are linked on the same chromosome. Regardless of whether the cause of genotypic correlation is pleiotropy or linkage, the fact is that genotypic correlations have important effects on response to multiple trait selection. Since genotypic correlations are functions of unobservable genotypic values, indirect methods are required for their estimation.

The standard definition of a correlation coefficient leads to the following definition of the genotypic correlation coefficient. The genotypic correlation between traits X and Y is the ratio of the genotypic covariance between the two traits to the geometric mean of their genotypic variances. That is,

$$r_G = \frac{\sigma_{G(XY)}}{[\sigma^2_{G(X)}\sigma^2_{G(Y)}]^{0.5}}$$

It has been shown that genotypic variances can be estimated either as the difference between phenotypic and environmental variances or as a function of the covariance between relatives. A similar pattern applies to the estimation of genotypic covariances.

Considering two traits, X and Y, let the phenotypic values be equal to the sum of genotypic values and environmental deviations. Thus,

$$P_{iX} = G_{iX} + E_{iX} \text{ and } P_{iY} = G_{iY} + E_{iY}$$

With proper randomization of genotypes with respect to environments, G_{iX} should not be correlated with E_{iX}, nor should G_{iY} be correlated with E_{iY}. G_{iX} and G_{iY} may be correlated due to pleiotropy or linkage and E_{iX} may be correlated with E_{iY} since both traits are subjected to the same environmental influences. The observable covariance between phenotypic values P_{iX} and P_{iY} will consist of

$$\sigma_{P(XY)} = \sigma_{[G(X) + E(X)][G(Y) + E(Y)]} = \sigma_{G(XY)} + \sigma_{E(XY)}$$

The phenotypic covariance is available directly from measurement of the two traits in a segregating or heterogeneous population. Estimates of environmental covariances will require repeated observations of both traits in homozygous or homogeneous populations. With segregating material, it will be necessary to include genetically homogeneous material (inbred parents, F_1s, or clones) to provide estimates of environmental covariances.

The second method for estimating genotypic covariances is by the use of covariances between relatives. In a random mating population, the phenotypic value of each trait can be considered to consist of an additive genetic value and an uncorrelated deviation. For the ith individual,

$$P_{iX} = G_{iX} + E_{iX} \text{ and } P_{iY} = G_{iY} + E_{iY}$$

where G_{iX} and G_{iY} now represent the additive genetic values of that individual. An offspring of the ith parent will have a phenotypic value for trait X, O_{iX}, consisting of $0.5\ G_{iX}$ contributed by that parent, a random genetic deviation contributed by the other parent, and a random environmental deviation. Since the latter two components will not be correlated with G_{iX}, G_{iY} or E_{iY}, the expected composition of the covariance between the phenotypic value for trait Y in the parent, P_{iY}, and the phenotypic value for trait X in the offspring, O_{iX}, will be

$$\sigma_{P(Y)O(X)} = \sigma_{[G(Y) + E(Y)][0.5\ G(X) + \text{ random deviation}]}$$

$$= 0.5\ \sigma_{G(XY)}$$

In a similar manner, $\sigma_{P(X)O(Y)} = 0.5\ \sigma_{G(XY)}$. To estimate the genotypic covariance, one would take twice the phenotypic covariance between one trait measured on the parent and the other trait measured on one of its random-bred progeny.

A related method for estimating the genotypic covariance is by relating response in one trait to selection for the other.

$$(G_{sY} - \overline{G}_Y) = b_{G(Y)P(X)} (P_{sX} - \overline{P}_X) = \frac{\sigma_{G(XY)}}{\sigma^2_{P(X)}} (P_{sX} - \overline{P}_X)$$

Thus, if selection is carried out for trait X and response measured in trait Y, the genotypic covariance can be estimated by dividing the observed response by the selection differential expressed as a proportion of the phenotypic variance of trait X, i.e.,

$$\sigma_{G(XY)} = (G_{sY} - \overline{G}_Y) \times \frac{\sigma^2_{P(X)}}{(P_{sX} - \overline{P}_X)}$$

This method is a special case of estimating genotypic covariances from covariances between relatives.

The discussion to this point serves to show the parallel between estimating genotypic covariances and estimating genotypic variances. In both cases, there are two basic methods to be considered. It should be possible to estimate genotypic variances in any experiment designed for estimating genotypic covariances. Furthermore, most experiments designed to estimate genotypic variances can give rise to estimates of genotypic covariances if both traits are measured on all individuals.

Once estimates of genotypic covariances and variances are available, the genotypic correlation can be estimated by applying the definition of the genotypic correlation given above. The equation for the genotypic correlation becomes quite complex when each of the three terms are expanded into its component parts. For example, if genotypic variances and covariances are estimated as differences between phenotypic and environmental variances and covariances, the genotypic correlation becomes

$$r_G = \frac{\sigma_{P(XY)} - \sigma_{E(XY)}}{[\sigma^2_{P(X)} - \sigma^2_{E(X)}]^{0.5} [\sigma^2_{P(Y)} - \sigma^2_{E(Y)}]^{0.5}}$$

The expression has become so complex that it is difficult to comprehend its meaning.

A useful approach to developing an understanding of the genotypic correlation can be derived from the formula, given by Hazel,[3] showing that the phenotypic correlation can be expressed as a function of the genotypic correlation, the environmental correlation, and the heritabilities of the two traits. According to Hazel,

$$r_P = h^2_X h^2_Y r_G + [1 - h^2_X]^{0.5} [1 - h^2_Y]^{0.5} r_E$$

where r_P, r_G, and r_E are the phenotypic, genotypic, and environmental correlations, respectively, and h^2_X and h^2_Y are the heritabilities of traits X and Y. From this equation,

$$r_G = \frac{r_P}{[h^2_X h^2_Y]^{0.5}} - \frac{[(1 - h^2_X)(1 - h^2_Y)]^{0.5} r_E}{[h^2_X h^2_Y]^{0.5}}$$

If the environmental correlation, r_E, is zero,

$$r_G = \frac{r_P}{[h^2_X h^2_Y]^{0.5}} = \frac{r_P}{h_X h_Y}$$

Since h^2_X and h^2_Y can never exceed 1.0, it is clear that the genotypic correlation will be larger in absolute magnitude than the phenotypic correlation, especially if either or both of the heritabilities are low. Furthermore, when $r_E = 0$, r_P can never exceed $[h^2_X h^2_Y]^{0.5}$ in absolute magnitude, even when the underlying genotypic correlation is 1.0 or -1.0. When the environmental correlation is not zero, the estimate of the genotypic correlation obtained by adjusting the phenotypic correlation for heritability must be further adjusted to take the environmental correlation into consideration.

While considering estimation of genotypic covariances from selection experiments, it was stated that response in trait Y to selection in trait X was a function of the genotypic covariance, the selection differential, and the phenotypic variance of trait X. This equation can be modified in the following way. Given that

$$(G_{sY} - \overline{G}_Y) = \frac{\sigma_{G(XY)}}{\sigma^2_{P(X)}} (P_{sX} - \overline{P}_X)$$

dividing both sides of the equation by the genotypic standard deviation of trait Y, and multiplying the numerator and the denominator of the right-hand side by the genotypic standard deviation of trait X, gives

$$\frac{(G_{sY} - \overline{G}_Y)}{\sigma_{G(Y)}} = r_G h_X \frac{(P_{sX} - \overline{P}_X)}{\sigma_{P(X)}}$$

Thus, expected correlated response in trait Y, when expressed in units of genotypic standard deviations, is equal to the product of (1) the genotypic correlation, (2) the heritability of the selected trait, and (3) the standardized selection differential. With this formulation, the central role of the genotypic correlation in correlated response to selection is emphasized.

V. SUMMARY

From the point of view of selection, whether it be for single or multiple traits, change in average genotypic value will depend upon the change in average phenotypic value and the degree of relationship between genotypic and phenotypic values. If genotypic and environmental effects act additively to produce the observed phenotypic response, the best estimate of genotypic value, or of change in average genotypic value, will be based on the principle of linear regression. If this requirement is met, the regression of genotypic values on phenotypic values is equal to the ratio of genotypic variance to phenotypic variance. Moreover, the coefficient of determination for linear regression of genotypic values on phenotypic values (or for linear regression of phenotypic values on genotypic values) is also equal to the ratio of genotypic variance to phenotypic variance (i.e., heritability).

Heritability and genotypic variances can be estimated indirectly from differences between phenotypic and environmental variances, or from the covariances between relatives. Selection experiments provide a special type of relationship between relatives which can also be used to estimate genotypic variance and heritability. Whether one should estimate total genotypic variance and broad-sense heritability, or additive genetic variance and narrow-sense heritability, depends on the particular application of the estimates. With inbred populations, broad-sense heritabilities may be more useful than narrow-sense heritabilities. The opposite is true when working with random mating populations.

Genotypic correlations may result from the effects of pleiotropy and/or linkage. The magnitude of the genotypic correlation between traits will have a very important impact on correlated response to selection. Many important decisions concerning the choice and application of selection indices depend on the magnitudes of the genotypic correlations among traits. Genotypic correlations and covariances, like heritabilities and genotypic variances, can only be estimated by indirect means. For this reason, they are difficult to measure with precision. Moreover, because of their complex nature, it may be difficult to develop an appreciation for their important impact on crop improvement efforts.

The relationships between the phenotypic, genotypic, and environmental correlations suggest that phenotypic correlations will often be lower in magnitude than their underlying genotypic correlations. This fact is important in trying to assess the strength of the relationships among traits in breeding program. Low heritability may mask quite strong genotypic relationships between some pairs of traits.

This brief introduction to some of the basic concepts in quantitative genetics should be helpful in understanding the basic principles of selection indices. An effort has been made to present these concepts in the framework of linear regression. This approach seems most useful for studying a range of problems which relate to artificial selection and crop improvement.

REFERENCES

1. **Comstock, R. E. and Moll, R. H.,** Genotype-environment interactions, in Statistical Genetics and Plant Breeding, No. 982, Hanson, W. D. and Robinson, H. F., Eds., National Academy of Sciences, National Research Council, Washington, D.C., 1963, 164.
2. **Nei, M.,** Studies on the application of biometrical genetics to plant breeding, *Mem. Coll. Agric. Kyoto Univ.,* 82, 1, 1960.
3. **Hazel, L. N.,** The genetic basis for constructing selection indexes, *Genetics,* 28, 476, 1943.

Chapter 4

METHODS FOR ESTIMATING AND EVALUATING GENOTYPIC AND PHENOTYPIC VARIANCES AND COVARIANCES

I. INTRODUCTION

Any experimental design that provides estimates of genotypic and phenotypic variances appropriate to single trait selection also will provide estimates of variances and covariances pertinent to the development and application of selection indices. In most applications, it is sufficient to restrict consideration to those experimental designs involving balanced data. In many instances, the experimental design will fit into the general design discussed by Comstock and Moll.[1] The choice of mating design used to produce genotypes for testing may often be based on the general considerations of Cockerham.[2] Hallauer and Miranda[3] have discussed specific designs for estimating genotypic variances and covariances in both cross- and self-pollinated species. Likewise, Mather and Jinks[4] have also discussed various methods for estimating genotypic variances in cross- and self-pollinated species.

The purpose of this chapter is to review the designs which are most apt to be used for estimating genotypic variances and covariances and, in some cases, to give examples of the calculations that are required to derive estimates and standard errors of those estimates. Since the precision of estimates is critical to their value in selection, some consideration is given to methods for evaluating estimates of variances and covariances.

II. MEASURES OF PRECISION OF ESTIMATES

It is possible to calculate approximate standard errors for estimates of variances and covariances. If M_{AA} and M_{BB} are mean squares for traits A and B, and M_{AB} is the mean cross-product, each measured with df degrees of freedom, the variance of M_{AA} is approximately $2M_{AA}^2/df$, the variance of M_{BB} is approximately $2M_{BB}^2/df$, and the variance of M_{AB} is approximately $(M_{AA}M_{BB} + M_{AB}^2)/df$ (see, for example, Tallis[5]). According to Kempthorne,[6] when calculating the variance of an estimated variance or covariance, one should replace expected mean squares and cross-products by their estimates and replace df by (df + 2). If an estimate of a variance or covariance is based on a linear function (F) of independent mean squares or mean cross-products, the following formula can be used to express the variance. If $F = \Sigma k_i M_i$, the variance of F is given by

$$\sigma_F^2 = \sum k_i^2 \, \sigma_{M(i)}^2$$

where $\sigma_{M(i)}^2$ is the variance of the ith mean square or cross-product. With these rules, it is possible to calculate an approximate standard error for any estimate of a variance or covariance, provided that the mean squares or mean cross-products included in the expression are independent. This will generally be assured in balanced designs.

Consider that the phenotypic variances of two traits, A and B, are estimated by the mean squares M_{AA} and M_{BB}, and that the phenotypic covariance between A and B is estimated by the mean cross-product M_{AB}, each with (m − 1) degrees of freedom. Further, consider that the environmental variances and covariance of traits A and B are estimated by m_{AA}. m_{BB} and m_{AB}, each measured with (n − 1) degrees of freedom. Then, the estimate of the genotypic variance of trait A would be $(M_{AA} − m_{AA})$ and, by the formula given above, would have an approximate variance of

$$\text{Var} \, (M_{AA} - m_{AA}) = \frac{2M_{AA}^2}{m + 1} + \frac{2m_{AA}^2}{n + 1}$$

Likewise, the genotypic covariance would be estimated by $(M_{AB} - m_{AB})$ with an approximate variance of

$$\text{Var } (M_{AB} - m_{AB}) = \frac{M_{AA}M_{BB} - M_{AB}^2}{m + 1} + \frac{m_{AA}m_{BB} - m_{AB}^2}{n + 1}$$

Heritability of trait A could be estimated by $(M_{AA} - m_{AA})/M_{AA}$. An approximate variance of the estimated heritability can be obtained by using the approximation to the variance of a ratio.[6] The ratio X/Y has an approximate variance of

$$\text{Var } \left(\frac{X}{Y}\right) = \frac{1}{Y^2} \left[\sigma_X^2 - 2 \frac{X}{Y} \sigma_{XY} + \frac{X^2}{Y^2} \sigma_Y^2\right]$$

Thus, if $X = (M_{AA} - m_{AA})$ and $Y = M_{AA}$, the variance of $X/Y = (M_{AA} - m_{AA})/M_{AA}$ is given by

$$\text{Var } \left[\frac{M_{AA} - m_{AA}}{M_{AA}}\right] = \frac{1}{M_{AA}^2} \left[\frac{2M_{AA}^2}{m + 1} + \frac{2m_{AA}^2}{n + 1}\right]$$
$$- \frac{2(M_{AA} - m_{AA})}{M_{AA}^3} \left[\frac{2M_{AA}^2}{m + 1}\right] + \frac{(M_{AA} - m_{AA})^2}{M_{AA}^4} \left[\frac{2M_{AA}^2}{m + 1}\right]$$

If expressed on a mean basis, heritability sometimes takes the form $h^2 = 1 - 1/F$, where F is the ratio of two mean squares. In these cases, the properties of the F-distribution can be used to calculate exact confidence intervals for heritability.[7]

The phenotypic correlation between traits A and B would be estimated by $r_P = M_{AB}/(M_{AA}M_{BB})^{0.5}$. In this case, the phenotypic correlation is the usual Pearson product-moment correlation with $(m - 2)$ degrees of freedom. Methods for testing the significance of such a correlation, and for developing confidence intervals, are presented in most statistical text books.

In some cases, the phenotypic correlation may be estimated as a ratio of linear functions of several mean squares and cross-products. This is always the case for genotypic correlations. Tallis[5] and Scheinberg[8] have presented methods for calculating approximate standard errors for these types of correlations.

More detailed examples of estimating variances or standard deviations of estimates of variance components and genotypic and phenotypic correlations are presented later in this chapter, particularly in sections VII and VIII.

III. ESTIMATION OF VARIANCES AND COVARIANCES FROM PARENT-OFFSPRING RELATIONSHIPS

Since response to selection depends upon the degree of relationship between phenotypic values of parents and those of their progeny, parent-offspring relationships are a natural source of information about genotypic and phenotypic variances and covariances. Estimates derived from parent-offspring relationships can be applied to selection among individuals equivalent to the parents, provided that response is to be assessed in the same type of offspring.

If, in a random mating population, both parents and progeny are represented by individual plants, the variance among parents provides an estimate of the phenotypic variance for any specified trait, while the covariance between two traits in the parents provides an estimate of the phenotypic covariance. For a single trait, the covariance between parents and their

Table 1
ANALYSIS OF COVARIANCE FOR BIPARENTAL PROGENIES
EVALUATED IN REPLICATED EXPERIMENTS

Source	Degrees of freedom[a]	Mean square or cross-product	Expectation of mean square or cross-product[b]
Replications	$r - 1$	$M_{1(vw)}$	$\sigma_{e(vw)} + pc\sigma_{r(vw)}$
Among crosses	$c - 1$	$M_{2(vw)}$	$\sigma_{e(vw)} + r\sigma_{w(vw)} + rp\sigma_{c(vw)}$
Within crosses	$c(p - 1)$	$M_{3(vw)}$	$\sigma_{e(vw)} + r\sigma_{w(vw)}$
Error	$(r - 1)(cp - 1)$	$M_{4(vw)}$	$\sigma_{e(vw)}$

[a] c = number of crosses, p = number of progeny per cross, r = number of clones per progeny each evaluated in one block of a randomized complete block design.

[b] $\sigma_{e(vw)}$ = environmental covariance among clones within genotypes, $\sigma_{w(vw)}$ = within-full-sib family genotypic covariance, $\sigma_{c(vw)}$ = genotypic covariance among full-sib family means, and $\sigma_{r(vw)}$ = environmental covariance among replication means.

respective progenies provides an estimate of one-half the additive genetic variance for that trait. Similarly, the covariance between one trait measured on the parents and another measured on the corresponding progenies provides an estimate of one-half the additive genetic covariance. In all cases, the estimates are applicable to questions concerning mass selection among single plants of one sex. In this case, expected response to mass selection for a single trait is proportional to one-half the additive genetic variance divided by the phenotypic variance. This is estimated by the covariance between parent and progeny divided by the variance among parents.

With inbred species, parents could be F_m-derived F_n lines while progeny are F_m-derived $F_{n'}$ lines ($n' > n$). Estimates of genotypic and phenotypic covariances in such a case would be pertinent to questions concerning selection among F_m-derived F_n lines where response is to be assessed in related F_m-derived $F_{n'}$ lines. For example, if $m = n = 2$ and $n' = 5$, estimates derived from parent-offspring relationships could be applied to selection among F_2 plants for average performance of F_2-derived F_5 lines. For a single trait, expected response would be proportional to the covariance between F_2 phenotypic values and the mean phenotypic value of the F_5 progeny divided by the phenotypic variance among F_2 plants.

IV. ESTIMATION OF VARIANCES AND COVARIANCES FROM BIPARENTAL PROGENY

One design which is sometimes used for estimating genotypic and phenotypic variances and covariances is simply to test the progeny of crosses between random pairs of parents from a given population. The progeny are sometimes referred to as biparental progeny, but may also be referred to as full-sib families. Parents are usually chosen from a random mating population and the progeny are usually evaluated as single plants. Hallauer and Miranda[3] discussed the analysis of a single trait for this design and concluded that it is not possible to separate the within-full-sib family genotypic variance from the plant-to-plant environmental variance. This limitation can be overcome in some species where individual plants can be cloned and subsequently evaluated in replicated tests.

An analysis of covariance for progeny derived from crosses between random pairs of parents is outlined in Table 1. The outline is based on the assumption that p progeny are produced in each of c crosses between random pairs of parents, and that each progeny plant can be cloned to give r propagules for evaluation. The experimental design is considered to be a randomized complete block design where one clone of each genotype is evaluated in each block. If the design were a completely random design, the first item would disappear from the table and the degrees of freedom for the error term would increase to $cp(r - 1)$. Expectations of mean cross-products for the remaining terms would be unchanged.

Most often, it will be difficult or impossible to clone individual plants. If so the first and fourth items in Table 1 will disappear from the analysis of covariance. The expectations of mean squares and mean cross-products of the remaining items will be as shown in Table 1 except that $r = 1$. It should be clear that, when $r = 1$, it will be impossible to separate $\sigma_{e(vw)}$ from $\sigma_{w(vw)}$.

For single traits, the genotypic variance among full-sib family means, $\sigma_{c(vv)} = \sigma_{c(v)}^2$ is known as the covariance between full-sibs. Similarly, $\sigma_{w(vv)} = \sigma_{w(v)}^2$ the genotypic variance among individuals within full-sib families, is equal to the total genotypic variance minus the covariance between full-sibs. These relationships allow one to relate this analysis to published genetic expectations of variances and covariances. In a random mating population, the covariance between full-sibs has a genetic expectation given by

$$\text{Cov FS} = \sigma_{c(vv)} = \frac{1}{2} V_A + \frac{1}{4} V_D + \frac{1}{4} V_{AA} + \frac{1}{8} V_{AD} + \frac{1}{16} V_{DD} + \ldots$$

where V_A = additive genetic variance, V_D = dominance genetic variance, V_{AA} = variance due to additive $\times$ additive epistatic interaction, V_{AD} = variance due to additive $\times$ dominance epistatic interaction, etc. The genetic expectation of the variance among plants within full-sib families is given by

$$V_G - \text{Cov FS} = \sigma_{w(vv)} = \frac{1}{2} V_A + \frac{3}{4} V_D + \frac{3}{4} V_{AA} + \frac{7}{8} V_{AD} + \ldots$$

Covariances between traits have similar expectations except that variances due to additive, dominance, and epistatic effects are replaced by covariances due to those effects.

If genotypic variances and covariances were the result of primarily additive gene action, $\sigma_{c(vw)}$ would be equal to one-half the additive genetic covariance. If one cannot assume that dominance and epistatic effects are unimportant, estimates of $\sigma_{c(vw)}$ would be biased estimates of one-half of the additive genetic covariance. In a random mating population, response to selection is dependent only upon the additive portion of the gene effects so that variances and covariances that are critical to selection response are those associated with additive effects. For this reason, $2\sigma_{c(vw)}$ is used as an estimate of the additive genetic covariance between traits v and w. When $v = w$, estimates of this component will be biased upwards as estimates of the additive genetic variance to the extent that dominance and epistatic effects are important. Additive genetic covariances will be estimated as

$$2\sigma_{c(vw)} = \frac{2}{rp} [M_{2(vw)} - M_{3(vw)}]$$

The phenotypic covariance for individual plants is equal to $\sigma_{c(vw)} + \sigma_{w(vw)} + \sigma_{e(vw)}/r'$, where r' is the number of clones to be used in evaluating each plant in the anticipated selection program. If $r = 1$ in the estimation experiment, it will not be possible to separate the environmental and genotypic causes of variation among plants and selection will therefore have to be planned on the basis of $r' = 1$. When $r = r' = 1$, the phenotypic covariance will be estimated by

$$\sigma_{P(vw)} = \frac{1}{p} [M_{2(vw)} - M_{3(vw)}] + M_{3(vw)}$$

When r > 1, the phenotypic covariance among progeny means of r' clones will be estimated by

$$\sigma_{P(vw)} = \frac{1}{rp} [M_{2(vw)} - M_{3(vw)}] + \frac{1}{r} [M_{3(vw)} - M_{4(vw)}] + \frac{1}{r'} M_{4(vw)}$$

If, rather than selecting on the basis of individual progeny values, one were interested in selection among full-sib family means based on p' sibs per family, the genotypic covariance would be estimated as indicated above, but would be multiplied by a different constant to predict response to selection. In such a case, the appropriate phenotypic covariance would be estimated by

$$\sigma_{P(vw)} = \frac{1}{rp} [M_{2(vw)} - M_{3(vw)}] + \frac{1}{rp'} [M_{3(vw)} - M_{4(vw)}] + \frac{1}{r'p'} M_{4(vw)}$$

Methods described in section II can be used to estimate genotypic and phenotypic correlations as well as heritabilities. As suggested by the above discussion, heritability estimates will be biased upward as estimates of narrow-sense heritability, because dominance and epistasis will cause an upward bias in the estimate of additive genetic variance.

Hallauer and Miranda[3] have pointed out that biparental progenies provide only a limited amount of information about genotypic and environmental variances. While the method is suitable for indicating if there is sufficient genotypic variability for selection, and whether or not there are strong associations between traits, there is concern that this method may not provide a sound basis for developing long-term selection strategies.

V. ESTIMATION OF VARIANCES AND COVARIANCES FROM DESIGN I EXPERIMENTS

Comstock and Robinson[9] described and evaluated three designs used at the North Carolina Experiment Station in efforts to determine the level of dominance of genes which affect quantitative traits. Although designed initially to be applied to F_2 populations of crosses between inbred lines, the designs have subsequently been applied to the estimation of variances and covariances in random mating populations.

In Design I experiments, plants are produced by crossing each of m male parents to a different set of n female parents to give mn full-sib families each with p progeny or full-sibs. Each plant can be classified into one of mn full-sib families and into one of m half-sib families. The design thus offers estimates based on both full-sib and half-sib family structure.

One possible experimental design for evaluating Design I material is to evaluate p plants from each full-sib family in each of r replications of a randomized complete block experiment. The analysis for such a design is outlined in Table 2. For a completely random design, the first item disappears and the degrees of freedom for experimental error becomes mn(r − 1). With one plant per plot (p = 1), the fifth item disappears from the analysis. The remaining items are unchanged except that p = 1.

In a random mating population, the genotypic variance among half-sib family means, $\sigma_{s(vv)} = \sigma_{s(v)}^2 = \text{Cov HS}$, is expected[10] to equal

$$\text{Cov HS} = \frac{1}{4} V_A + \frac{1}{16} V_{AA} + \frac{1}{64} V_{AAA} + \dots$$

where V_A is the additive genetic variance, V_{AA} is the variance due to additive × additive

Table 2

ANALYSIS OF COVARIANCE FOR DESIGN I EXPERIMENTS

Source	Degrees of freedom[a]	Mean square or cross-product	Expectation of mean square or cross-product[b]
Replications	r − 1	$M_{1(vw)}$	$\sigma_{e(vw)} + p\sigma_{e(vw)} + mnp\sigma_{r(vw)}$
Among male parents	m − 1	$M_{2(vw)}$	$\sigma_{w(vw)} + p\sigma_{e(vw)} + pr\sigma_{d(vw)} + prn\sigma_{s(vw)}$
Among female parents within males	m(n − 1)	$M_{3(vw)}$	$\sigma_{w(vw)} + p\sigma_{e(vw)} + pr\sigma_{d(vw)}$
Error	(r − 1) (mn − 1)	$M_{4(vw)}$	$\sigma_{w(vw)} + p\sigma_{e(vw)}$
Among plants within plots	mnr(p − 1)	$M_{5(vw)}$	$\sigma_{w(vw)}$

a   p = number plants per plot, r = number of replications, m = number of males (half-sib families), and n = number of females mated to each male (number full-sib families in each half-sib family).

b $\sigma_{w(vw)}$ = phenotypic covariance among plants within plots.

      $\sigma_{e(vw)}$ = environmental covariance among plots within replications,

      $\sigma_{d(vw)}$ = genotypic covariance among full-sib family means,

      $\sigma_{s(vw)}$ = genotypic covariance among half-sib family means, and

      $\sigma_{r(vw)}$ = environmental covariance among replication means.

epistatic interaction, etc. The average variance, within half-sib families, of full-sib family means, $\sigma_{d(vv)} = \sigma^2_{d(v)} = \text{Cov FS} - \text{Cov HS}$, has the following genetic expectation:

$$\text{Cov FS} - \text{Cov HS} = \frac{1}{4} V_A + \frac{1}{4} V_D + \frac{3}{16} V_{AA} + \frac{1}{8} V_{AD} + \frac{1}{16} V_{DD} + \ldots$$

Based on these expectations, it is clear that $\sigma_{s(vv)}$ is composed primarily of variance due to additive effects while $\sigma_{d(vv)}$ should exceed $\sigma_{s(vv)}$ by an amount that will be dependent primarily on dominance and epistatic effects. Since selection response depends upon the additive genetic variation and covariation, genotypic variances and covariances are usually estimated by

$$4\sigma_{s(vw)} = \frac{4}{prn} [M_{2(vw)} - M_{3(vw)}]$$

An estimate of the effect of dominance can be obtained by estimating $4[\sigma_{d(vw)} - \sigma_{s(vw)}]$. However, this component is not required for problems of index selection.

The definitions of phenotypic variances and covariances will depend upon the anticipated basis of selection. If mass selection is to be carried out, one should estimate the phenotypic variance and covariances for individual plants. An appropriate estimate is given by

$$\sigma_{P(vw)} = \sigma_{w(vw)} + \sigma_{e(vw)} + \sigma_{d(vw)} + \sigma_{s(vw)}$$

$$= M_{5(vw)} + \frac{1}{p} [M_{4(vw)} - M_{5(vw)}]$$

$$+ \frac{1}{pr} [M_{3(vw)} - M_{4(vw)}] + \frac{1}{prn} [M_{2(vw)} - M_{3(vw)}]$$

For selection among full-sib family means, the definitions of both genotypic and of phenotypic variances and covariances must be changed. The appropriate genotypic variances and covariances would be estimated by

$$2\sigma_{s(vw)} = \frac{2}{prn} [M_{2(vw)} - M_{3(vw)}]$$

The corresponding phenotypic variances and covariances, if each full-sib family is to contain p' plants in each of r' plots, would be

$$\sigma_{P(vw)} = \sigma_{s(vw)} + \sigma_{d(vw)} + \frac{1}{r'} \sigma_{e(vw)} + \frac{1}{p'r'} \sigma_{w(vw)}$$

$$= \frac{1}{prn} [M_{2(vw)} - M_{3(vw)}] + \frac{1}{pr} [M_{3(vw)} - M_{4(vw)}]$$

$$+ \frac{1}{pr'} [M_{4(vw)} - M_{5(vw)}] + \frac{1}{p'r'} M_{5(vw)}$$

$$= \frac{1}{prn} M_{2(vw)} + \frac{n-1}{prn} M_{3(vw)} + \frac{r-r'}{prr'} M_{4(vw)} + \frac{p-p'}{pp'r'} M_{5(vw)}$$

Similar adjustments would be required for selection based on half-sib family means or for selection within full- or half-sib families.

In the above discussion, it has been assumed that the parents are randomly sampled from a random mating population and then randomly mated to produce full-sib families. Because the design was originally developed for use in an F_2 of a cross between homozygous inbred lines, there may be interest in trying to apply the method in cases where the parents are randomly chosen from a partially inbred population. The discussion to this point will apply equally to parents being a random sample from an F_2, provided that linkage does not play a role in the inheritance of any of the traits under consideration. If linkage is important, the genetic expectations of the various variances and covariances become complex functions of gene effects and linkage parameters. For a random mating population, one can argue that the population is in equilibrium. In applying Design I methods to an F_2 population, one must limit use of the derived statistics to plans relating to selection within the F_2 population. The estimates derived from the analysis would not apply to any other generation in the selfing series because they would not reflect the same degree of inbreeding.

If the parents used in Design I crossing schemes are inbred, estimates of parameters can be obtained by adjusting the expectations given in Table 2 for the inbreeding coefficient.[2] It is necessary to assume that all full-sib families have reached the same level of inbreeding. Even in this case, however, the estimated variances and covariances can be applied only to questions of selection among noninbred individuals. To obtain estimates that could be applied to selection among partially inbred lines, it is necessary to inbreed the full-sib families to the same level before evaluating them in a modified Design I analysis. Inbreeding is often accompanied by inadvertant selection and this will further complicate interpretation of Design I analysis of partially inbred populations.

VI. ESTIMATION OF VARIANCES AND COVARIANCES FROM DESIGN II EXPERIMENTS

The second design described by Comstock and Robinson[9] has also been used to estimate genotypic parameters in random mating populataions.[3] With multiflowered species, it is possible to cross m males in all combinations with n females to produce mn full-sib families. Individuals may be classified according to full-sib family, according to half-sib family sharing a common male parent, and according to half-sib family sharing a common female parent.

Since the number of crosses in which one parent, particularly the female parent, can be

Table 3
**ANALYSIS OF COVARIANCE FOR DESIGN II EXPERIMENTS REPLICATED
IN SEVERAL ENVIRONMENTS**

Source	Degrees of freedom[a]	Mean square or cross-product	Expectation of mean square or cross-product[b]
Environments (E)	$t - 1$		
Sets (S)	$g - 1$		
E × S	$(t - 1)(g - 1)$		
Reps in E × S	$gt(r - 1)$		
Males in S (M)	$g(m - 1)$	$M_{1(vw)}$	$\sigma_{w(vw)} + p\sigma_{e(vw)} + pr\sigma_{asd(vw)} + prn\sigma_{as(vw)} + prt\sigma_{sd(vw)} + prtn\sigma_{s(vw)}$
Females in S (F)	$g(n - 1)$	$M_{2(vw)}$	$\sigma_{w(vw)} + p\sigma_{e(vw)} + pr\sigma_{asd(vw)} + prm\sigma_{ad(vw)} + prt\sigma_{sd(vw)} + prtm\sigma_{d(vw)}$
M × F	$g(m - 1)(n - 1)$	$M_{3(vw)}$	$\sigma_{w(vw)} + p\sigma_{e(vw)} + pr\sigma_{asd(vw)} + prt\sigma_{sd(vw)}$
E × M	$g(t - 1)(m - 1)$	$M_{4(vw)}$	$\sigma_{w(vw)} + p\sigma_{e(vw)} + pr\sigma_{asd(vw)} + prn\sigma_{as(vw)}$
E × F	$g(t - 1)(n - 1)$	$M_{5(vw)}$	$\sigma_{w(vw)} + p\sigma_{e(vw)} + pr\sigma_{asd(vw)} + prm\sigma_{ad(vw)}$
E × M × F	$g(t - 1)(m - 1)(n - 1)$	$M_{6(vw)}$	$\sigma_{w(vw)} + p\sigma_{e(vw)} + pr\sigma_{asd(vw)}$
Pooled error	$gt(r - 1)(mn - 1)$	$M_{7(vw)}$	$\sigma_{w(vw)} + p\sigma_{e(vw)}$
Sampling error	$rgtmn(p - 1)$	$M_{8(vw)}$	$\sigma_{w(vw)}$

[a] g = number of sets of m males crossed with n females,
      r = number of replications in each of t environments, and
      p = number of plants per plot.

[b] $\sigma_{w(vw)}$ = phenotypic covariance among plants within plots,
      $\sigma_{e(vw)}$ = error covariance between plots,
      $\sigma_{asd(vw)}$ = covariance due to interactions among environmental(a), male(s), and female(d) effects,
      $\sigma_{ad(vw)}$ = covariance due to interaction of environmental and female effects,
      $\sigma_{as(vw)}$ = covariance due to interaction of environmental and male effects,
      $\sigma_{sd(vw)}$ = covariance due to interaction between male and female effects,
      $\sigma_{d(vw)}$ = covariance due to female effects, and
      $\sigma_{s(vw)}$ = covariance due to male effects.

used is usually quite limited, several sets of mn crosses need to be produced.[9] The analysis of Design II data consists of separating variation among full-sib family means into portions attributable to differences among half-sib family means where half-sibs share the same female parent, differences among half-sib family means where half-sibs share the same male parent, and a residual or interaction between male and female parents. These estimates are derived for each of s sets and pooled over sets.

Various experimental designs may be used to test material developed from a Design II mating scheme. Comstock and Robinson[9] suggested that each set be tested in a replicated experiment. They based their analysis on data measured on a per plot or plot mean basis. If sufficient seed can be produced from each mating, the entire experiment might be repeated in two or more environments. If one anticipated that selection would be based on individual plants, it would be useful to collect data on individual plants within each plot. An outline of an analysis of covariance for Design II material tested in several environments and with data collected on several plants within each plot is given in Table 3.

If the Design II material is tested in a single environment, means squares and cross-products including $M_{4(vw)}$, $M_{5(vw)}$, and $M_{6(vw)}$ disappear from the analysis given in Table 3. In this case, the usual estimates of $\sigma_{d(vw)}$, $\sigma_{s(vw)}$, and $\sigma_{sd(vw)}$ become biased by the inclusion of effects due to genotype-environment interaction. If data are collected on a plot basis only, item $M_{8(vw)}$ cannot be estimated. The remaining expectations are unchanged except that p = 1. When p = 1, covariation due to plant-to-plant phenotypic differences within plots cannot be separated from the covariation due to plot-to-plot environmental differences.

For individual traits, the variance components for genotypic differences among male and among female half-sib family means, $\sigma_{s(vw)}$ and $\sigma_{d(vw)}$, are both equal to the covariance between half-sibs[10] and have the following genotypic expectations in a random mating population:

$$\sigma_{s(vv)} = \sigma_{d(vv)} = \frac{1}{4} V_A + \frac{1}{16} V_{AA} + \frac{1}{64} V_{AAA} + \ldots$$

In this equation, V_A is the additive genetic variance for trait v in the population, V_{AA} is the variance due to additive $\times$ additive types of epistasis, and so on. As in Design I, an estimate of the additive genetic covariance in the random mating reference population is given by

$$4\sigma_{s(vw)} = \frac{4}{prtn} [M_{1(vw)} - M_{3(vw)} - M_{4(vw)} + M_{6(vw)}]$$

or by

$$4\sigma_{d(vw)} = \frac{4}{prtm} [M_{2(vw)} - M_{3(vw)} - M_{5(vw)} + M_{6(vw)}]$$

Since both components have the same expectation, it would be best to use the average of the two components to estimate genotypic covariances. If the number of males in each set is equal to the number of females in each set (m = n), one could pool $M_{1(vw)}$ and $M_{2(vw)}$, as well as $M_{4(vw)}$ and $M_{5(vw)}$, to obtain one estimate of the covariance among half-sib family means and one estimate of the environment $\times$ half-sib family interaction. Equivalently, one could use $2\sigma_{s(vw)} + 2\sigma_{d(vw)}$ as the estimate of the additive genetic covariance. When numbers of females and males within each set are not equal, the first approach would give estimates of genotypic covariances with somewhat smaller standard errors than the equal weighted estimate.

The mean square for males $\times$ females within sets ($M_{3(vw)}$ in Table 3) includes a component, $\sigma_{sd(vv)}$, which measures the residual variation among full-sib family means after first correcting for the average contributions of each of the two parents. This is comparable to the subdivision of variance in a two-way table into sources relating to the main effects (based on marginal totals) and a residual or interaction variance. Since the expected genotypic variance among full-sib families is

$$Cov\ FS = \frac{1}{2} V_A + \frac{1}{4} V_D + \frac{1}{4} V_{AA} + \frac{1}{8} V_{AD} + \frac{1}{16} V_{DD} + \frac{1}{8} V_{AAA} + \ldots$$

and the variance among male half-sib family means plus the variance among female half-sib family means is expected to be

$$2\ Cov\ HS = \frac{1}{2} V_A + \frac{1}{8} V_{AA} + \frac{1}{32} V_{AAA} + \ldots$$

the component $\sigma_{sd(vv)}$ is expected to equal

$$\text{Cov FS} - 2 \text{ Cov HS} = \frac{1}{4} V_D + \frac{1}{8} V_{AA} + \frac{1}{8} V_{AD} + \frac{1}{16} V_{DD}$$

$$+ \frac{3}{32} V_{AAA} + \ldots$$

with similar expressions for the covariances between traits. If epistasis is not important, covariances due to dominant gene effects can be estimated by $4\sigma_{sd(vw)}$. From Table 3,

$$4\sigma_{sd(vw)} = \frac{4}{prt} [M_{3(vw)} - M_{6(vw)}]$$

As with Design I, estimates of genotypic and phenotypic variances and covariances will vary according to the intended selection program. For mass selection, genotypic covariances should be estimated by the additive genetic covariance while the phenotypic covariances should reflect all items which affect plant-to-plant covariation. For family selection based on half-sib or full-sib family means, the appropriate genotypic covariance should be an estimate of one-quarter or one-half of the additive genetic covariance, respectively. The phenotypic covariance must be an estimate of the variance of half-sib or full-sib family means evaluated in a defined sample of replications and/or environments. Consideration of the following two scenarios should suffice to demonstrate the types of calculations that might be required.

For mass selection, the additive genetic covariance can be estimated by

$$\sigma_{g(vw)} = 2\sigma_{s(vw)} + 2\sigma_{d(vw)}$$

$$= \frac{2}{prtn} [M_{1(vw)} - M_{3(vw)} - M_{4(vw)} + M_{6(vw)}]$$

$$+ \frac{2}{prtm} [M_{2(vw)} - M_{3(vw)} - M_{5(vw)} + M_{6(vw)}]$$

The total phenotypic covariance among individual plants would be estimated by

$$\sigma_{P(vw)} = \sigma_{s(vw)} + \sigma_{d(vw)} + \sigma_{sd(vw)} + \sigma_{as(vw)} + \sigma_{ad(vw)}$$

$$+ \sigma_{asd(vw)} + \sigma_{e(vw)} + \sigma_{w(vw)}$$

$$= \frac{1}{prtn} [M_{1(vw)} - M_{3(vw)} - M_{4(vw)} + M_{6(vw)}]$$

$$+ \frac{1}{prtm} [M_{2(vw)} - M_{3(vw)} - M_{5(vw)} + M_{6(vw)}]$$

$$+ \frac{1}{prt} [M_{3(vw)} - M_{6(vw)}] + \frac{1}{prn} [M_{4(vw)} - M_{6(vw)}]$$

$$+ \frac{1}{prm} [M_{5(vw)} - M_{6(vw)}] + \frac{1}{pr} [M_{6(vw)} - M_{7(vw)}]$$

$$+ \frac{1}{p} [M_{7(vw)} - M_{8(vw)}] + M_{8(vw)}$$

Methods described in section II can be used to calculate standard errors for the covariance estimates as well as for calculating genotypic and phenotypic correlations and heritabilities.

For full-sib family selection, expected response to single trait selection is dependent upon one-half the additive genetic variance divided by the standard deviation of full-sib family means.[3] Thus, genotypic covariances for this type of selection should be estimated by 2 Cov HS which, with equal weighting of the male and female within sets cross-products, becomes

$$\sigma_{g(vw)} = \sigma_{s(vw)} + \sigma_{d(vw)}$$

$$= \frac{1}{prtn} [M_{1(vw)} - M_{3(vw)} - M_{4(vw)} + M_{6(vw)}]$$

$$+ \frac{1}{prtm} [M_{2(vw)} - M_{3(vw)} - M_{5(vw)} + M_{6(vw)}]$$

The phenotypic covariance among full-sib family means each based on the evaluation of p′ plants in each of r′ plots in each of t′ environments would be

$$\sigma_{P(vw)} = \sigma_{s(vw)} + \sigma_{d(vw)} + \sigma_{sd(vw)} + \frac{1}{t'} \sigma_{as(vw)} + \frac{1}{t'} \sigma_{ad(vw)}$$

$$+ \frac{1}{t'} \sigma_{asd(vw)} + \frac{1}{r't'} \sigma_{e(vw)} + \frac{1}{p'r't'} \sigma_{w(vw)}$$

$$= \frac{1}{prtn} M_{1(vw)} + \frac{1}{prtm} M_{2(vw)} + \frac{mn - m - n}{prtmn} M_{3(vw)} + \frac{t - t'}{prtt'n} M_{4(vw)}$$

$$+ \frac{t - t'}{prtt'm} M_{5(vw)} + \frac{t - t'}{prtt'mn} M_{6(vw)} + \frac{r - r'}{prr't'} M_{7(vw)} + \frac{p - p'}{pp'r't'} M_{8(vw)}$$

If r′ = r, p′ = p and t′ = t,

$$\sigma_{P(vw)} = \frac{m}{prtmn} M_{1(vw)} + \frac{n}{prtmn} M_{2(vw)} + \frac{mn - m - n}{prtmn} M_{3(vw)}$$

Heritability of full-sib family means for trait v would be estimated as $h_v^2 = \sigma_{g(vv)}/\sigma_{P(vv)}$, the genotypic correlation between traits v and w would be estimated as

$$r_{g(vw)} = \frac{\sigma_{g(vw)}}{[\sigma_{g(vv)}\sigma_{g(ww)}]^{0.5}}$$

and the phenotypic correlation between traits v and w would be estimated by

$$r_{P(vw)} = \frac{\sigma_{P(vw)}}{[\sigma_{P(vv)}\sigma_{P(ww)}]^{0.5}}$$

Table 4 contains essential elements from the analysis of simulated data for a Design II experiment. For demonstration purposes, it was assumed that 20 sets of 24 full-sib families, developed by crossing each of four males with each of six females within each set, were evaluated in three plots at each of two locations. It was considered that two traits had been measured and that data had been recorded on a plot basis in each case.

Essential mean squares and mean cross-products were calculated from the simulated data and are presented in Table 4 along with estimates of the critical components of covariances and their standard errors. The estimated components were then combined to give estimates

Table 4
ANALYSIS OF COVARIANCE FOR SIMULATED DATA FROM A DESIGN II EXPERIMENT REPLICATED IN TWO ENVIRONMENTS

Source	Degrees of freedom[a]	Mean square or cross-product		
		A × A	A × B	B × B
Males in sets (M)	60	2332.7	− 186.65	172.08
Females in sets (F)	100	1562.9	− 123.20	110.32
M × F	300	276.5	− 4.70	3.98
M × environments	60	566.8	− 14.16	4.24
F × environments	100	439.9	− 12.00	5.86
M × F × envir.	300	291.3	− 8.66	4.02
Pooled error	1840	240.9	− 6.39	4.15

Covariance components[b]	Estimates		
	A × A	A × B	B × B
$\sigma_{s(vw)}$	49.5(12.0)	− 4.9(2.3)	4.66(0.86)
$\sigma_{d(vw)}$	47.4(9.6)	− 4.8(1.8)	4.35(0.65)
$\sigma_{sd(vw)}$	− 2.5(5.5)	0.7(0.5)	− 0.01(0.08)
$\sigma_{as(vw)}$	15.3(5.8)	− 0.3(0.4)	0.01(0.05)
$\sigma_{ad(vw)}$	12.4(5.5)	− 0.3(0.5)	0.15(0.07)
$\sigma_{asd(vw)}$	16.8(8.3)	− 0.8(0.7)	− 0.04(0.12)
$\sigma_{e(vw)}$	240.9(7.9)	− 6.4(0.8)	4.15(0.14)

[a]   Degrees of freedom based on 20 sets of 24 full-sib families (4 males × 6 females), evaluated in three replications in each of two environments.
[b]   See Table 3 and text for definition of variance components.

of the genotypic and phenotypic variances and covariances required for the application of index selection to selection among full-sib family means where each mean is to be based on three replications ($r' = r = 3$) in each of two environments ($t' = t = 2$). For trait A, the genotypic variance was estimated as 96.9 with a standard error of 15.4, and the phenotypic variance among full-sib family means as 156.8 with a standard error of 15.0. This gives an estimated heritability for trait A of 96.9/156.8 = 0.62 = 62%. For trait B, the estimated genotypic variance was 9.02 with a standard error of 1.07, while the phenotypic variance was estimated as 9.76 with a standard error of 1.07. This gave an estimate of the heritability of trait B of 9.02/9.76 = 0.92 = 92%. The genotypic covariance was estimated to be − 9.7 with a standard error of 3.0 and the phenotypic covariance was − 10.8 with a standard error of 2.9. The genotypic correlation, $-9.7/[(96.9)(9.02)]^{0.5} = -0.33$, and the phenotypic correlation, $-10.8/[(156.8)(9.76)]^{0.5} = -0.28$, were both negative.

The discussion of Design II has been based on the assumption that the parents were a random sample from a random mating population. In this way, relevant genetic covariances could be interpreted in terms of the genetic characteristics of a random mating reference population. A random mating population should be in linkage equilibrium, and the complexities caused by linkage do not need to enter into a consideration of genetic constitution of covariances.

Design II was developed originally to be applied to the F_2 population of a cross between two inbred parents. In order to determine the average level of dominance,[9] it was important to know that gene frequencies at segregating loci were 0.5. If there is no linkage between genes affecting various quantitative traits, the F_2 population will be equivalent to a random mating population in terms of gene and genotype frequencies. Since the design was developed for progeny of a cross between inbred lines, one might ask if the design can be used for estimating variances and covariances which could be used for developing selection indices for use in inbred species.

Cockerham[2] discussed the consequences of inbreeding in the parents of a design such as Design II, and pointed out that inbreeding does affect the genetic expectations of the components listed in Table 3. If the parents are inbred, but mated at random, the statistics from a Design II analysis are applicable to a random mating population with the same gene frequencies. This is because the progenies evaluated in the Design II experiment are no longer inbred by virtue of the random mating among parents. However, inbreeding of the parental generation does require that the inbreeding coefficient (assumed equal for all parents, or at least for all full-sib families) be taken into consideration in estimating the variances and covariances of the random mating reference population. For example, if the average level of inbreeding in the parents of a Design II analysis is F, the half-sib covariance, estimated from the male or female sources of variation, has a genetic expectation of

$$\text{Cov HS} = \frac{1 + F}{4} V_A + \left[\frac{1 + F}{4}\right]^2 V_{AA} + \ldots$$

This requires that estimates of additive genetic covariances be taken as $4\sigma_{s(vw)}/(1 + F)$ or $4\sigma_{d(vw)}/(1 + F)$, and so on. Cockerham[2] suggested that to estimate components that are to be used in planning selection in partially inbred populations, it is necessary to inbreed the full-sib families from a Design II mating to the same degree as anticipated in the selection program prior to evaluation and estimation of the covariances.

VII. ESTIMATION OF VARIANCES AND COVARIANCES FROM INBRED OR PARTIALLY INBRED LINES

In breeding programs for self-pollinated species, breeding methods such as the modified pedigree method[11] require selecting among partially inbred lines. To develop a selection index for multiple trait selection in such a program requires estimates of variances and covariances among a random sample of such lines. Plant breeders are often concerned about average performance over a sample of environments, perhaps over several locations or years. The experimental design for such a test is that described for a single trait by Comstock and Moll.[1] The statistical model for the phenotypic value of the vth trait measured on the ith genotype tested in the qth replication at the jth location in the kth year is given by

$$P_{ijkq(v)} = \mu_{(v)} + a_{j(v)} + b_{k(v)} + (ab)_{jk(v)} + r_{jkq(v)}$$
$$+ g_{i(v)} + (ga)_{ij(v)} + (gb)_{ik(v)} + (gab)_{ijk(v)} + e_{ijkq(v)}$$

where μ is the overall mean, g refers to genotypic value, a refers to location effect, b refers to year effect, and e refers to plot-to-plot environmental deviations. In this model, it is assumed that the genotypic effects, $g_{i(v)}$, are random, as are the interactions between genotypic effects and location or year effects.

Comstock and Moll[1] explained how the definition of phenotypic value will depend upon the intended selection program. If selection is to be based on mean performance over r' replications in each of s' locations for t' years, the phenotypic value of interest is the average for each genotype over those environments; i.e.,

$$P_{i...(v)} = \frac{1}{r's't'} \sum\sum\sum P_{ijkq(v)}$$

Table 5
ANALYSIS OF COVARIANCE FOR INBRED LINES EVALUATED IN REPLICATED EXPERIMENTS AT SEVERAL LOCATIONS IN SEVERAL YEARS

Source	Degrees of freedom[a]	Mean cross-product	Expectation of mean cross-product[b]
Locations (L)	$s - 1$		
Years (Y)	$t - 1$		
L × Y	$(s - 1)(t - 1)$		
Reps in Y × L	$st(r - 1)$		
Genotypes (G)	$n - 1$	$M_{1(vw)}$	$\sigma_{e(vw)} + r\sigma_{gab(vw)} + rt\sigma_{ga(vw)} + rs\sigma_{gb(vw)} + rst\sigma_{g(vw)}$
G × L	$(n - 1)(s - 1)$	$M_{2(vw)}$	$\sigma_{e(vw)} + r\sigma_{gab(vw)} + rt\sigma_{ga(vw)}$
G × Y	$(n - 1)(t - 1)$	$M_{3(vw)}$	$\sigma_{e(vw)} + r\sigma_{gab(vw)} + rs\sigma_{gb(vw)}$
G × Y × L	$(n - 1)(s - 1)(t - 1)$	$M_{4(vw)}$	$\sigma_{e(vw)} + r\sigma_{gab(vw)}$
Pooled error	$st(r - 1)(n - 1)$	$M_{5(vw)}$	$\sigma_{e(vw)}$

[a] n = number genotypes evaluated in r replications at s locations in each of t years.

[b] Subscripts g, a, and b refer to the effects of genotype, location, and year, respectively. Subscript e refers to the environmental effect associated with an experimental unit (plot).

The phenotypic variances and covariances that need to be estimated are those which are appropriate for this phenotypic mean. The phenotypic variances and covariances will be linear functions of the variances and covariances of $g_{i(v)}$, $(ga)_{ij(v)}$, $(gb)_{ik(v)}$, $(gab)_{ijk(v)}$ and $e_{ijkq(v)}$ effects, and will vary depending upon the proposed selection program. In fact, the variance of $P_i \ldots (v)$ will be

$$\sigma_{P(vv)} = \sigma_{g(vv)} + \frac{1}{s'} \sigma_{ga(vv)} + \frac{1}{t'} \sigma_{gb(vv)} + \frac{1}{s't'} \sigma_{gab(vv)}$$
$$+ \frac{1}{r's't'} \sigma_{e(vv)}$$

where the variances represent the variances of the genotypic effects, of the genotype-location interaction effects, of the genotype-year interaction effects, of the genotype-location-year interaction effects, and of the environmental effects associated with individual experimental units. Note that $a_{j(v)}$, $b_{k(v)}$, $(ab)_{jk(v)}$, and $r_{jkq(v)}$ are constant for all genotypes and, therefore, do not enter into the variances of mean phenotypic values. The covariance between phenotypic mean values $P_i \ldots (v)$ and $P_i \ldots (w)$ is likewise a function of the corresponding covariance terms; i.e.,

$$\sigma_{P(vw)} = \sigma_{g(vw)} + \frac{1}{s'} \sigma_{ga(vw)} + \frac{1}{t'} \sigma_{gb(vw)} + \frac{1}{s't'} \sigma_{gab(vw)}$$
$$+ \frac{1}{r's't'} \sigma_{e(vw)}$$

where the terms now refer to the covariances between the corresponding effects for traits v and w. Each variance and covariance in the expression given above can be estimated as a linear function of mean squares or mean cross-products calculated from an analysis of covariance.

The analysis of covariance for this type of experiment is outlined in Table 5 with expectations of mean squares and mean cross-products being given only to those items which appear in estimates of genotypic and phenotypic variances and covariances. In an experiment where n genotypes are evaluated in r replications at each of s locations in t years, the genotypic covariances between traits v and w (variances if v = w) are estimated by

$$\sigma_{g(vw)} = \frac{1}{rst} [M_{1(vw)} - M_{2(vw)} - M_{3(vw)} + M_{4(vw)}]$$

Similarly, the estimates of the other covariance components are

$$\sigma_{ga(vw)} = \frac{1}{rt} [M_{2(vw)} - M_{4(vw)}]$$

$$\sigma_{gb(vw)} = \frac{1}{rs} [M_{3(vw)} - M_{4(vw)}]$$

$$\sigma_{gab(vw)} = \frac{1}{r} [M_{4(vw)} - M_{5(vw)}] \text{ and}$$

$$\sigma_{e(vw)} = M_{5(vw)}$$

where $M_{i(vw)}$ is the ith mean cross-product if v is not equal to w and the ith mean square if v = w. Similarly, $\sigma_{g(vw)}$ is the genotypic covariance if v is not equal to w and the genotypic variance, sometimes written as $\sigma^2_{g(v)}$ if v = w.

If a selection program is to be conducted using r' replications at s' locations in t' years, the phenotypic covariance for genotypic means would be given by

$$\sigma_{P(vw)} = \sigma_{g(vw)} + \frac{1}{s'} \sigma_{ga(vw)} + \frac{1}{t'} \sigma_{gb(vw)} + \frac{1}{s't'} \sigma_{gab(vw)}$$

$$+ \frac{1}{r's't'} \sigma_{e(vw)}$$

which can be estimated by

$$\sigma_{P(vw)} = \frac{1}{rst} [M_{1(vw)} - M_{2(vw)} - M_{3(vw)} + M_{4(vw)}]$$

$$+ \frac{1}{rs't} [M_{2(vw)} - M_{4(vw)}] + \frac{1}{rst'} [M_{3(vw)} - M_{4(vw)}]$$

$$+ \frac{1}{rs't'} [M_{4(vw)} - M_{5(vw)}] + \frac{1}{r's't'} M_{5(vw)}$$

$$= \frac{1}{rst} M_{1(vw)} + \frac{s - s'}{rs'st} M_{2(vw)} + \frac{t - t'}{rst't} M_{3(vw)} + \frac{(s - s')(t - t')}{rss'tt'} M_{4(vw)}$$

If the genotypic means are based on the same number of replications, locations, and years as used in estimating the variances and covariances (i.e., r' = r, s' = s, and t' = t), then the phenotypic covariance simplifies to $\sigma_{P(vw)} = M_{1(vw)}/rst$. The genotypic covariance will always be estimated by

$$\sigma_{g(vw)} = \frac{1}{rst} [M_{1(vw)} - M_{2(vw)} - M_{3(vw)} + M_{4(vw)}]$$

regardless of the anticipated environmental sampling in the selection program.

When the experiment for estimating variances and covariances is carried out in a single year (t = 1), the mean squares and mean cross-products for years (Y), genotype-year interaction (G × Y), and genotype-location-year interaction (G × L × Y) cannot be estimated. For this reason, it is not possible in these cases to obtain separate estimates of $\sigma_{g(vw)}$ and $\sigma_{gb(vw)}$, nor of $\sigma_{ga(vw)}$ and $\sigma_{gab(vw)}$. Likewise, if the experiment is carried out at only one location (s = 1), it will not be possible to obtain separate estimates of $\sigma_{g(vw)}$ and $\sigma_{ga(vw)}$, nor of $\sigma_{gb(vw)}$ and $\sigma_{gab(vw)}$. With s = 1 and t = 1, the estimated genotypic covariance becomes an estimate of $\sigma_{g(vw)} + \sigma_{ga(vw)} + \sigma_{gb(vw)} + \sigma_{gab(vw)}$. In this case, the estimates may be useful only if one can safely assume that one or more of the confounded components is sufficiently small to be ignored. Estimates based on experiments at one location may be useful if selection is to be practiced at that location and if one is concerned only with response at that location. Then, any effects of genotype-location interaction can properly be considered as part of the total genotypic effect at that location and any observed interaction among genotype, location, and year can properly be considered as a component of the genotype-year interaction at that particular location.

Computer simulation was used to generate data suitable for demonstrating the analysis outlined above. For this simulation, it was assumed that the genotypic values for each trait were normally distributed and that the values for the two traits were correlated. Year and location effects were considered to be fixed effects, while those for genotype, and all interactions with genotype, were considered to be random normally distributed effects with mean equal to zero. For this simulation, the correlation between traits was considered to be the same for all effects in the model. The simulation was designed to generate data for 30 genotypes evaluated in four replications at each of three locations in each of two years; a total of 720 observations on each of two traits. Program BACOVAI (Appendix 5) was used to calculate the required mean squares and mean cross-products (Table 6).

It is possible to test the mean squares for each trait for significant deviation from the hypothesis that the corresponding variance component is zero. For genotypes, one must use an approximate F-test.[5] For trait A, the F-value for genotypes is (954557 + 92334)/(192760 + 126767) = 3.28. The approximate degrees of freedom associated with the numerator is $(954557 + 92334)^2/(954557^2/29 + 92334^2/58) = 35$ while the denominator degrees of freedom is taken as $(192760 + 126767)^2/(192760^2/58 + 126767^2/29) = 85$. Under the null hypothesis that the genotypic component of variance is zero for trait A, the probability of an F-ratio with 35 and 85 degrees of freedom being as large as 3.28 is less than 0.01, or 1%. One would therefore conclude that there is a significant genotypic variance for trait A in this example.

Further study of the expectations of mean squares (Table 5) shows that both the genotype-location and the genotype-year mean squares should be tested against the genotype-location-year mean square. For trait A, the F-ratio for the genotype-location interaction is 192760/92334 = 2.09 with 58 and 58 degrees of freedom. Since the value exceeds the tabular value for probability equal to 0.01, one must conclude that there is significant variability in trait A due to interaction between genotype and location effects. The F-ratio for the genotype-year interaction is 126767/92334 = 1.37 with 29 and 58 degrees of freedom. This is smaller than the tabular value for probability equal to 0.05 and degrees of freedom of 29 and 58. Thus, there is insufficient reason for rejecting the hypothesis that all genotype-year interaction effects are zero and that the corresponding variance component is, therefore, equal to zero.

The genotype-location-year interaction mean square should be compared to the pooled error mean square. For trait A, the F-ratio is 92334/41863 = 2.21 which, with 38 and 552

Table 6
ANALYSIS OF COVARIANCE FOR SIMULATED DATA FOR
TWO TRAITS MEASURED ON 30 INBRED LINES EVALUATED
IN FOUR REPLICATIONS AT EACH OF THREE LOCATIONS IN
EACH OF TWO YEARS

Source	Degrees of freedom	Mean square or cross-product		
		A × A	A × B	B × B
Genotypes	29	954557	4288.4	93.359
Genotypes × locations	58	192760	249.8	3.321
Genotypes × years	29	126767	235.0	4.412
Genotypes × locations × years	58	92334	90.7	1.593
Pooled error	522	41863	53.9	0.840

Covariance components[a]	Estimates		
	A × A	A × B	B × B
$\sigma_{g(vw)}$	30307(10319)	162.3(78.0)	3.634(0.990)
$\sigma_{ga(vw)}$	12553(4878)	19.9(14.9)	0.216(0.084)
$\sigma_{gb(vw)}$	2869(3029)	12.0(12.5)	0.235(0.096)
$\sigma_{gab(vw)}$	12618(4264)	9.2(12.9)	0.188(0.074)
$\sigma_{e(vw)}$	41863(2586)	53.9(8.5)	0.840(0.052)

[a] Standard error in brackets.

degrees of freedom, is significant at a probability of less than 0.01. The pooled error mean square is an estimate of the variance of the environmental effects associated with individual experimental units (plots). The analysis of variance does not provide any test of this mean square. However, one must always recognize that there will be a nonzero variation among experimental units.

For trait A, F-ratios indicate that there were significant variances for genotypic effects, for genotype-location interactions, and for the three-way (second order) interactions among genotype, location, and year. The genotype-year interaction was judged to be not significantly different from zero. Similar tests for trait B indicated that all sources of variation listed in Table 6 were significant.

The components of variance and covariance listed in Table 6 were calculated in the manner described above and their standard errors were estimated by the methods outlined in section II. As an example of the calculations, consider the estimation of the genotypic covariance between trait A and B as well as the estimation of its standard error.

$$\sigma_{g(AB)} = [4288.4 - 249.8 - 235.0 + 90.7]/24 = 162.3$$

The variance of this estimate is given by

$$([(954557)(93.359) + 4288.4^2]/31 + [(192760)(3.321) + 249.8^2]/60 + [(126767)(4.412) + 235.0^2]/31 + [(92334)(1.593) + 90.7^2]/60)/24^2 = 6084.0$$

Thus, the standard error of the estimated genotypic component of covariance is $6084.0^{0.5} = 78.0$.

The standard errors of estimated variance and covariance components cannot be used for tests of significance of those components because the true distributions of those estimates are not known. However, reports in the literature often make use of a rule of thumb which is based on the assumption that the underlying distribution is normal. If the underlying distribution of sample variances and covariances was normal, estimates which exceed their standard error by a factor of 1.96 would be considered significant at a probability level of 5%. That is, if an estimate exceeds 1.96 times its standard error, one can be 95% confident that the corresponding population value is greater than zero. For this reason, when estimated components of variance or covariance exceed twice their standard error, one usually concludes that the true component of variance or covariance is different from zero. In the example (Table 6), this rule of thumb would lead to the same conclusions as did the F-tests of the mean squares. Application of this rule to the estimated components of covariance (Table 6) leads to the conclusion that only the genotypic and error components of covariance can be considered significantly different from zero.

There is a danger of misinterpretation when applying the above rule or when testing mean squares by F-tests. If a component is judged to be nonsignificant by either method, many researchers are inclined to conclude that the null hypothesis is true. That is, they conclude that the component of variance or covariance is equal to zero. In fact, the component is likely to be different from zero but the experiment was not large enough to show significance. Chew[13] discussed the benefits of estimating parameters along with a confidence interval rather than testing hypotheses about the parameter. In the present example, it is more realistic to conclude that the true variance of genotype-year interaction effects for trait A may be as low as zero or as high as $2869 + 2(3029) = 8927$, than to conclude that it is equal to zero.

In this type of experimentation, it is not unusual to find that estimates of variance components have negative values. This will happen whenever the corresponding F-ratio is less than 1.0. Possible approaches to handling negative estimates are discussed in a later section of this chapter.

Estimated components of variance and covariance (Table 6) can be combined to give estimates of phenotypic variances and covariances appropriate to any anticipated selection program. If one were anticipating selection to be based on mean performance over two replications ($r' = 2$) at each of three locations ($s' = 3$) in a single year ($t' = 1$), the phenotypic variance or covariance would be

$$\sigma_{P(vw)} = \sigma_{g(vw)} + \frac{1}{3}\sigma_{ga(vw)} + \sigma_{gb(vw)} + \frac{1}{3}\sigma_{gab(vw)} + \frac{1}{6}\sigma_{e(vw)}$$

For trait A, the phenotypic variance would be estimated by

$$(30307 + 12553/3 + 2869 + 12618/3 + 41863/6) = 48544$$

For this type of selection, the estimated heritability for trait A would be $30307/48544 = 0.62 = 62\%$. For trait B, the equivalent phenotypic variance is 4.144 which gives an estimate of the heritability of trait B of $3.634/4.144 = 0.88 = 88\%$. The equivalent phenotypic covariance is 193.0 giving an estimate of the phenotypic correlation of $193.0/[(48544)(4.144)]^{0.5} = 0.43$. The estimated genotypic correlation would be $162.3/[(30307)(3.634)]^{0.5} = 0.49$.

If one were concerned about selection based on four replications at each of three locations in two years, the estimated phenotypic variances for traits A and B would be 39773 and 3.890 with a covariance of 178.7. This gives estimates of heritability for trait A of 76% and for trait B of 93%, as well as an estimate of 0.45 for the phenotypic correlation. Heritabilities based on the more extensive evaluation program are higher, as expected.

It is possible to calculate an approximate standard error for heritability estimates based on linear functions of mean squares. For testing based on four replications, three locations,

Table 7

OUTLINE OF ANALYSIS OF VARIANCE OR COVARIANCE FOR m PLOTS (p PLANTS PER PLOT) OF GENETICALLY HOMOGENEOUS PLANTS AND n PLOTS (q PLANTS PER PLOT) OF GENETICALLY HETEROGENEOUS PLANTS

Source of variation	Degrees of freedom	Expectation of mean square or mean cross-product[a]
Homogeneous plots		
Among plots	$m - 1$	$\sigma_{e(vw)} + p\sigma_{a(vw)}$
Among plants within plots	$m(p - 1)$	$\sigma_{e(vw)}$
Heterogeneous plots		
Among plots	$n - 1$	$\sigma_{P(vw)} + \sigma_{b(vw)}$
Among plants within plots	$n(q - 1)$	$\sigma_{P(vw)}$

> [a] $\sigma_{P(vw)}$ and $\sigma_{e(vw)}$ are the phenotypic and environmental covariances (or variances if v = w) between traits v and w based on deviations between observations on individual plants and true plot means; $\sigma_{a(vw)}$ is the covariance for true means of plots of homogeneous material and will include a factor for differences between means of different homogeneous populations; $\sigma_{b(vw)}$ is the covariance between true phenotypic means of plots of heterogeneous material.

and two years, the method described by Pesek and Baker[14] gives an estimate of heritability for trait A of 76% with a standard error of 8% and for trait B of 93% with a standard error of 2%. Use of the conservative method of Dickerson[15] gives corresponding standard errors of 26% for heritability of trait A and 25% for the heritability of trait B.

VIII. ESTIMATION OF VARIANCES AND COVARIANCES FROM COMPARISONS OF HETEROGENEOUS AND HOMOGENEOUS POPULATIONS

In self-pollinated species, phenotypic variation and covariation in genotypically heterogeneous populations (F_2, F_3, F_4, etc.) can be compared to the average variation and covariation within one or more genotypically homogeneous populations (parents, F_1) to provide estimates of total genotypic variances and covariances. While these experiments may be conducted by completely randomizing individual plants of the various populations, a more common design is to group the plants into single rows or small plots and to have replications of the rows or plots of each type of population. In the latter case, genotypic and phenotypic variances and covariances should be estimated from the within-plot deviations, and the estimates should be applied to questions concerning selection among plants within plots. Estimates are based on pooled values where pooling is over all plots of a particular population.

A typical type of experiment is one in which m rows or plots of homogeneous populations and n rows or plots of a heterogeneous population are planted and data are collected on p plants in each of the homogeneous rows and q plants in each of the heterogeneous rows. The homogeneous rows might include m/4 of each of the two inbred parents and m/2 of the F_1 of the cross between them. The heterogeneous population could be the F_2 represented by q plants in each of n rows or plots. The analysis of this type of experiment would follow the outline given in Table 7. In this design, the phenotypic variances and covariances among plants within rows are estimated directly from the mean squares and mean cross-products for the ''among plants within plots'' of heterogeneous populations. Environmental variances and covariances are estimated from the mean squares and mean cross-products of the ''among plants within plots'' of the homogeneous populations. The genotypic variances and covar-

Table 8
ANALYSIS OF SIMULATED DATA FOR TWO TRAITS MEASURED
IN HOMOGENEOUS AND HETEROGENEOUS POPULATIONS

Source of variation	Degrees of freedom	Mean square or cross-product		
		Trait A	A × B	Trait B
Homogeneous populations (P_1, P_2, F_1)				
Among plots	39	67.105	− 64.490	66.097
Among plants within plots	160	0.681	0.111	0.610
Within P_1	40	0.701	0.285	0.794
Within P_2	40	0.487	0.209	0.392
Within F_1	80	0.769	− 0.024	0.626
Heterogeneous population (F_2)				
Among plots	39	6.626	− 1.929	6.866
Among plants within plots	160	6.525	− 1.262	3.582

iances are then estimated by subtracting the environmental variances and covariances from the corresponding phenotypic variances and covariances.

In some cases, researchers choose to give different weights to estimates of environmental variances and covariances obtained from different homogeneous populations. For example, if the heterogeneous population is an F_2, consideration of the frequencies of the three genotypes at a particular locus leads to the suggestion that the estimates of environmental parameters should be based on 1:2:1 weightings of the estimates derived from one parent, the F_1, and the other parent. The analysis outlined in Table 7 will give such a weighting, provided that there is a 1:2:1 proportion of plots of the corresponding homogeneous populations. If not, estimates would need to be developed separately for each population and then combined to give a weighted average.

A frequent observation in this type of analysis is that different estimates of environmental variances are not homogeneous. This indicates a type of genotype-environment interaction in which the effects of genotype and environment are not additive in their determination of phenotypic expression. If this occurs, one should search for a transformation which, when applied to the data, will equalize the estimates of environmental variance. Without equal variances, it is difficult to argue that the difference between the phenotypic variance and the estimate of environmental variance (based on some average of the different environmental variances) is truly an estimate of the genotypic variance.

An example of the type of analysis outlined in Table 7 is given in Table 8. The data were simulated on a microcomputer using a program like that in Appendix 2. Simulation was of a 15-locus model where each locus carried alleles with equal additive effects. Plus alleles at loci 1-7 increased trait A and had no effect on trait B, plus alleles at loci 8-10 increased trait A and decreased trait B, and plus alleles at loci 11-15 had no effect on trait A but increased trait B. All loci were considered to be unlinked. The design consisted of five plants per plot with 10 plots of each of two inbred parents, 20 plots of the F_1, and 40 plots of the F_2.

From the analysis given in Table 8, one can estimate the phenotypic variances for traits A and B directly as the mean squares for "among plants with plots" of the heterogeneous F_2 population. Thus, the phenotypic variance for trait A is 6.525 while that for trait B is 3.582. Similarly, the phenotypic covariance between traits A and B is − 1.262. The genotypic parameters are estimated as differences between mean squares or mean cross-products. For trait A, the genotypic variance is estimated as the difference between the variance of the F_2 and the variance for the homogeneous populations, i.e., 6.525 − 0.681 = 5.844. Similarly, the genotypic variance for trait B is estimated as 3.582 − 0.610 = 2.972. The estimate of the genotypic covariance between the two traits is $-1.262 - 0.111 = -1.373$. From these estimates, one would calculate the genotypic correlation between traits A and B, based on within-plot deviations, as the genotypic covariance between A and B divided by the product of the genotypic standard deviations of the two traits, i.e., $-1.373/[(5.844)(2.972)]^{0.5}$ $= -0.33$. Likewise, the phenotypic correlation is -0.26. The estimates of the heritabilities of the within-plot deviations are $5.844/6.525 = 0.90 = 90\%$ for trait A and $2.972/3.582$ $= 0.83 = 83\%$ for trait B.

In practice, one should use Bartlett's test[12] to test the assumption of homogeneity of environmental variances for each of the two traits. For trait A, the chi-squared statistic for the test of homogeneity of the estimates from the three homogeneous populations (0.701, 0.487, and 0.769) is 2.66 which, with 2 degrees of freedom, fails to indicate any significant departure from homogeneity ($P > 0.25$). For trait B, the chi-squared value of 4.99 also fails to indicate any significant departure from the assumption of homogeneous environmental variances ($P > 0.05$).

One must also assume that the environmental covariances estimated from the three homogeneous populations are homogeneous. There does not seem to be a comparable test for the homogeneity of estimated environmental covariances. One might, however, estimate the environmental correlations for each of the three homogeneous populations and test them for homogeneity. The test that is usually applied is based on the fact that, for large sample sizes, the hyperbolic tangent of the correlation coefficient is approximately normally distributed with variance $1/(n - 3)$ where n is the number of pairs of observations. In an analysis of variance context, it would seem that $(n - 3)$ should be replaced by the degrees of freedom associated with the mean squares and cross-products, less 2. If this calculation is carried out for the three sample correlations (0.382 for P_1, 0.478 for P_2, and -0.035 for F_1), the chi-squared value is found to be 7.71 ($0.01 < P < 0.025$) which suggests that the environmental correlation, and therefore the environmental covariance, is lower in the F_1 than in either of the parents.

The rules described in section II of this chapter can be applied to derive standard errors for the estimates of phenotypic and genotypic variances and covariances. The variance of the estimated phenotypic variance of trait A is $2(6.525)^2/162 = 0.526$ so that the standard error of the estimated phenotypic variance (6.525) is $0.526^{0.5} = 0.725$. For trait B, the estimated phenotypic variance is 3.582 with a standard error of 0.398. The variance of the estimated phenotypic covariance is $[(6.525)(3.582) + 1.262^2]/162 = 0.154$ so that the estimated phenotypic covariance is -1.262 with a standard error of 0.392.

The calculations for the standard errors of estimates of genotypic parameters are somewhat more complex. For trait A, the variance of the estimated genotypic variance is

$$2(0.681)^2/162 + 2(6.525)^2/162 = 0.531$$

so that the estimate of the genotypic variance for trait A is 5.844 with a standard error of $0.531^{0.5} = 0.729$. Similarly, for trait B, the estimated genotypic variance is 2.972 with a standard error of 0.404. The variance of the estimated genotypic covariance is

$$[(0.681)(0.610) + 0.111^2]/162 + [(6.525)(3.582) + 1.262^2]/162 = 0.157$$

Thus, the estimated genotypic covariance is -1.373 with a standard error of 0.396.

Calculating standard errors for estimates of genotypic and phenotypic variances and covariances is not required for selection indices. Moreover, since the distributions of the sample estimates of variances and covariances are not known, these estimates cannot be used to test hypotheses or develop confidence intervals for the population parameters. However, routine calculation of the standard errors will provide some additional information about the precision of estimates. Observation of very large standard errors for some estimates may motivate a researcher to carry out a more thorough assessment of their value in selection indices.

In using comparisons of homogeneous and heterogeneous populations to develop estimates of phenotypic and genotypic variances and covariances, one might ask what is the best allocation of resources to testing the two types of population. Since the standard error for the estimated genotypic parameters will always be greater than for the corresponding phenotypic parameters, it seems reasonable to try to design the experiment so that the standard errors of the genotypic parameters will be as small as possible with the resources available. Consider that plants from both types of population are to be grown in a completely randomized experiment where the phenotypic variance will be estimated by M_P, a mean square with kR $- 1$ degrees of freedom, and where the environmental variance will be estimated by M_e, a mean square with $(1 - k)R - 1$ degrees of freedom. Then, for a given total number of plants (R) to be tested, the objective is to choose k, the proportion of plants to be from the heterogeneous population, so as to minimize the standard error of the estimate of genotypic variance. This can be accomplished by minimizing the variance of $(M_P - M_e)$, where, by the rules described in section II,

$$\text{Var}\ (M_P - M_e) = \frac{2M_P^2}{kR + 1} + \frac{2M_e^2}{(1 - k)R + 1}$$

Single plant heritability, h^2, is defined as $(M_P - M_e)/M_P$ so that $M_e = (1 - h^2)M_P$. Incorporating this identity, rearranging, and ignoring the $+1$ in the denominator, the above expression becomes

$$\text{Var}\ (M_P - M_e) = \frac{2M_P^2}{R} \left[\frac{1}{k} + \frac{(1 - h^2)^2}{1 - k} \right]$$

where the expression in square brackets is to be minimized with respect to k. Differentiating and setting the derivative to zero shows that the expression will be minimum if $k = 1/(2 - h^2)$. Thus, using an h^2 approximately equal to h^2 of the trait with the lowest heritability, one would allocate a proportion k of the total plants to be tested to the heterogeneous population and proportion $(1 - k)$ to the homogeneous populations. The proportion to be allocated to the heterogeneous population will vary from 0.5, when heritability is close to zero, to 1.0 when heritability approaches 1.0. For intermediate heritabilities ($h^2 = 0.3$ to 0.5), approximately two-thirds of the resources should be allocated to testing the hetero-geneous population and one-third to testing the homogeneous populations.

Variances and covariances estimated by this method can be used in relation to selection in the population from which the heterogeneous sample was taken. Use of these estimates in other populations requires justification for assuming that the other populations have the same variance-covariance structure.

The method gives estimates of genotypic parameters which are based on total genotypic effects rather than additive genotypic effects (breeding values). If this method is applied to early generations, the estimates of variances and covariances will be biased upward by dominance and dominant types of epistasis, and may not adequately reflect the variance and covariances of the breeding values of individuals in the population. This criticism should not be so serious if the method is applied to later generations (say F_4 or F_5) where genotypic values of individual plants are more highly correlated with the average genotypic values of their progeny.[16] The estimates are usually derived from evaluation of single plant perform-ance. When genotypes can be replicated, the method outlined in the last section may be more appropriate.

IX. INSPECTION OF ESTIMATED GENOTYPIC AND PHENOTYPIC COVARIANCE MATRICES

Hayes and Hill[17] recommended that estimated genotypic and phenotypic covariance matrices be subjected to a transformation which would better show their inherent properties. The transformation results in new variables which have phenotypic variances of 1.0, phenotypic covariances of 0.0, genotypic variances equal to the eigenvalues, **k**, of the determinantal equation [$Gx = kPx$], and genotypic covariances of 0.0. Thus, if any of the eigenvalues have negative values, there is an indication that, on the untransformed scale, some of the partial genotypic correlations exceed the acceptable bounds of -1 to $+1$. Hayes and Hill pointed out that if any of the eigenvalues exceed 1.0, there is an indication that some of the partial environmental correlations exceed the bounds of -1 to $+1$. In either case, there is some question as to whether or not the estimated covariance matrices will give rise to reliable estimates for selection indices.

The method proposed by Hayes and Hill[17] can be demonstrated by use of the variance-covariance matrices estimated in the Design II simulation experiment of section VI. For the two traits considered in that example, the phenotypic variances and covariances were

$$P = \begin{bmatrix} 156.8 & -10.8 \\ -10.8 & 9.76 \end{bmatrix}$$

and the genotypic variances and covariances were

$$G = \begin{bmatrix} 96.9 & -9.7 \\ -9.7 & 9.02 \end{bmatrix}$$

The method first requires the solution of the determinantal equation which, for two variables, is quadratic. This equation,

$$(96.9 - 156.8k)(9.02 - 9.76k) - (-10.8 + 9.7k)(-10.8 + 9.7k) = 0$$

has to be solved for the eigenvalues k. While this solution is really all that is required for evaluation of the variance-covariance matrices, it is sometimes instructive to estimate the associated eigenvectors.

In this example, the smallest of the two eigenvalues was estimated to be $k_1 = 0.5968$ with associated eigenvector $x' = (0.084608 \ 0.322108)$. The larger eigenvalue was $k_2 = 0.9244$ with eigenvector $x' = (0.083067 \ 0.001898)$. These calculations indicate that the original variables A and B may be transformed as follows:

$$W = 0.083067 \ A + 0.084608 \ B \text{ and}$$

$$Y = 0.001898 \ A + 0.322108 \ B$$

The new variables, W and Y, have the following properties. The phenotypic variance of W is

$$(0.083067)^2(156.8) + (2)(0.083067)(0.084608)(-10.8) + (0.084608)^2(9.76) = 1.0$$

Similarly, the phenotypic variance of Y is

$$(0.001898)^2(156.8) + (2)(0.001898)(0.322108)(-10.8) + (0.322108)^2(9.76) = 1.0$$

The phenotypic covariance between W and Y is

$$(0.083067)(0.001898)(156.8) + (0.083067)(0.322108)(-10.8) +$$
$$(0.084608)(0.001898)(-10.8) + (0.084608)(0.322108)(9.76) = 0.0$$

Replacing the phenotypic variance and covariances in the three expressions given above by their genotypic counterparts shows that the genotypic variance of the transformed variables W and Y are 0.5968 and 0.9244, respectively, and that the genotypic covariance between W and Y is 0.0.

Since both eigenvalues lie in the range of 0 to 1, there is no evidence to suggest that the estimates are not reasonable estimates of the corresponding population parameters. On this basis, one might be fairly confident in using them to develop selection indices. Hayes and Hill[17] suggested that selection indices might be quite unreliable if the average of the eigenvalues was close to zero.

One of the prerequisites of applying the method of Hayes and Hill[17] is that the phenotypic matrix of variances and covariances be positive-definite; i.e., that it not have any roots which are negative or zero. When the phenotypic variances and covariances are estimated as linear functions of two or more sets of mean squares and mean cross-products, it is possible that this requirement is not met. A program (Appendix 9) was written to develop the Hayes-Hill transformation. That program will check that the phenotypic covariance matrix is positive-definite and will not proceed unless that requirement is met. If the phenotypic matrix is non-positive-definite, removal of one or more of a set of highly correlated variables will often correct the problem. For example, the phenotypic variance-covariance matrix derived from Tables 2, 3, and 4 of Robinson et al.[18] is not positive-definite. It was noted that the variable "husk score" was equal to the variable "husk length" plus a measure of husk tightness. Husk score and husk length were closely associated with a phenotypic correlation of 0.96. After removing husk score from the variance-covariance matrix, it became positive-definite. The eigenvalues for the remaining seven variables were 0.0374, 0.1710, 0.2272, 0.2684, 0.3677, 0.3723, and 0.4721, indicating that the estimated matrices should provide a reasonable basis for developing selection indices.

Yousaf[19] presented variance-covariance matrices for five traits in maize. Estimates were based on a Design II experiment. Calculation of the eigenvalues for the Hayes-Hill[17] transformation gave values of 0.1219, 0.3190, 0.4897, 0.8782, and 18.1130. The value of 18.1130 far exceeds 1.0 and suggests that at least one of the partial environmental correlations exceeds the acceptable range of -1 to $+1$. This result raises questions about how useful the estimated variances and covariances are. In such instances, it may be useful to calculate the environmental correlations and partial correlations to see where the problem lies. Perhaps the problem can be corrected by removing a variable which has been measured with low precision. In this particular instance, the environmental correlation between number of rows of kernels and 100 kernel weight was -7.42 suggesting a typographical error in the published table.[19] It seems that the negative sign was omitted for the genetic covariance component for these two variables. When corrected, the new eigenvalues were 0.1276, 0.3887, 0.8530, 0.8872, and 1.7049. Calculation of the first order partial environmental correlations still failed to explain the remaining large eigenvalue. However, several of the higher order partial correlations were outside the acceptable range indicating that sampling variability had resulted in imprecise estimates of variances and covariances. In this case, there is no objective rule for whether or not the estimates should be abandoned. Present practice would suggest at least that the researcher be aware of the imprecision in the estimates and recognize that such imprecision will result in indices which are far less efficient than an optimum index.

X. MAXIMUM LIKELIHOOD ESTIMATION OF VARIANCE COMPONENTS

The usual approach to estimating variance components is to equate expectations of mean squares to observed mean squares, and to solve the resulting set of equations. This has been the procedure discussed in this chapter, but it requires discussion in light of recent research into using maximum likelihood methods for estimating variance components.

Estimates of variance components will sometimes be negative. Negative estimates present a problem in interpretation since they are estimates of parameters which must be positive by definition. Searle[20] reviewed possible strategies for coping with negative estimates of variance components.

Estimates may be negative because of insufficient data or because the wrong model has been used in their estimation. One strategy for handling negative estimates is to accept them at face value. An alternative is to consider that they indicate that the true value of the corresponding component is zero, and to replace the negative estimate with a value of zero. This approach introduces a bias into the estimation procedure.

Thompson[21] developed a procedure, known as restricted maximum likelihood, for developing nonnegative estimates of variance components. Essentially, the method consists of removing variance components for which negative estimates have been obtained from the model, and reestimating the remaining components. This approach seems to be a better approach than merely replacing negative estimates with zero. If all estimates are positive, the restricted maximum likelihood estimates are identical to those developed by analysis of variance as described in this chapter.

One possible explanation for the occurrence of negative estimates is that the method used for their estimation is weak. It is this possibility that has led to the development of maximum likelihood methods for estimating variance components. In maximum likelihood methods, maximization should be over the nonnegative parameter space and should therefore lead to nonnegative estimates of all variance components. Corbeil and Searle[22] developed maximum likelihood estimators for variance components in one-way and two-way designs. They concluded that maximum likelihood estimates would be less variable than estimates from analysis of variance in most, but not all, situations.

The differences among estimates derived from analysis of variance, restricted maximum likelihood, and maximum likelihood procedures can be seen by reference to a one-way, random effects model. If the design consists of g groups each with n observations, and if MSA represents the among-group mean square with $(g - 1)$ degrees of freedom while MSE represents the within-group mean square with $g(n - 1)$ degrees of freedom, the analysis of variance estimates for the variance components are

$$\sigma_A^2 = \frac{1}{n} [\text{MSA} - \text{MSE}] \text{ and } \sigma_E^2 = \text{MSE}$$

The restricted maximum likelihood estimates[21] are identical if MSA is greater than or equal to MSE. If MSA is less than MSE, the estimates are given by

$$\sigma_A^2 = 0 \text{ , and } \sigma_E^2 = \frac{g - 1}{ng - 1} \text{MSA} + \frac{g(n - 1)}{gn - 1} \text{MSE}$$

For the unrestricted maximum likelihood estimates, Corbeil and Searle[22] indicated that

$$\sigma_A^2 = \frac{1}{n}\left[\frac{g-1}{g}\, MSA - MSE\right] \text{ and } \sigma_E^2 = MSE$$

In this case, and in any others for which maximum likelihood estimators have been developed, the methods differ only in the coefficients which are used to combine mean squares into estimates of variance components. The mean squares are calculated in the usual way.

Maximum likelihood estimators for variance components have been developed for only a few designs and are not generally available, even for balanced designs. Moreover, there is still some disagreement over what constitutes optimum properties for estimators of variance components. Recently, Lee and Kapadia[23] questioned some of the earlier results and conclusions. They favored the restricted maximum likelihood method of Thompson[21] to the unrestricted maximum likelihood method of Corbeil and Searle.[22]

Because maximum likelihood estimators of variance components have not been developed for all balanced designs, and because there is a lack of general agreement as to what constitutes the best estimation procedure, the method used in this chapter has been that based on equating observed mean squares and cross-products to their expectations. It is known that this method gives unbiased estimates of variance components. If the estimates are positive, it is also true that they are equal to the restricted maximum likelihood estimates of Thompson.[21] Should future research establish that other estimates are superior, it is likely that those new estimates will consist of minor modifications to the linear functions of mean squares and cross-products developed in this chapter. With such an eventuality, it will not be difficult to adjust the methods developed in this chapter to accomodate a better estimation procedure.

XI. SUMMARY

The most popular methods for estimating genotypic and phenotypic variances and covariances have been reviewed. Worked examples have been given for three of the methods. The worked examples include methodology for calculating approximate standard errors for estimates of variances and covariances, and for estimates of heritability. Researchers are urged to routinely calculate standard errors as a caution against using estimates which are grossly unreliable.

A section has been included to demonstrate a recent suggestion for testing the adequacy of estimates of variances and covariances. This new method consists of developing a transformation for the data which greatly simplifies the interpretation of the variance-covariance matrices. While the method does not include a rule for when estimates are reliable enough to be used in developing selection indices, it does provide a powerful tool for detecting imprecise estimates.

REFERENCES

1. **Comstock, R. E. and Moll, R. H.,** Genotype-environment interactions, in Statistical Genetics and Plant Breeding, No. 982, Hanson, W. D. and Robinson, H. F., Eds., National Academy of Science, National Research Council, Washington, D.C., 1963, 164.
2. **Cockerham, C. C.,** Estimation of genetic variances, in Statistical Genetics and Plant Breeding, No. 982, Hanson, W. D. and Robinson, H. F., Eds., National Academy of Sciences, National Research Council, Washington, D.C., 1963, 53.
3. **Hallauer, A. R. and Miranda, J. B., Fo.,** *Quantitative Genetics in Maize Breeding,* Iowa State University Press, Ames, 1981.
4. **Mather, K. and Jinks, J. L.,** *Biometrical Genetics. The Study of Continuous Variation,* Chapman and Hall, London, 1971.

5. **Tallis, G. M.,** Sampling errors of genetic correlation coefficients calculated from analyses of variance and covariance, *Aust. J. Stat.,* 1, 35, 1965.

6. **Kempthorne, O.,** *An Introduction to Genetic Statistics,* John Wiley & Sons, New York, 1957.

7. **Knapp, S. J., Stroup, W. W., and Ross, W. M.,** Exact confidence intervals for heritability on a progeny mean basis, *Crop Sci.,* 25, 192, 1985.

8. **Scheinberg, E.,** The sampling variance of the correlation coefficients estimated in genetic experiments, *Biometrics,* 22, 187, 1966.

9. **Comstock, R. E. and Robinson, H. F.,** Estimation of average dominance of genes, in *Heterosis,* Gowan, J. W., Ed., Iowa State University Press, Ames, 1952, 494.

10. **Falconer, D. S.,** *Introduction to Quantitative Genetics,* Longman, London, 1981.

11. **Brim, C. A.,** A modified pedigree method of selection in soybeans, *Crop Sci.,* 6, 220, 1966.

12. **Cochran, W. G. and Cox, G. M.,** *Experimental Designs,* John Wiley & Sons, New York, 1957.

13. **Chew, V.,** Statistical hypothesis testing: An academic exercise in futility, *Proc. Fla. State Hortic. Soc.,* 90, 214, 1977.

14. **Pesek, J. and Baker, R. J.,** Comparison of predicted and observed responses to selection for yield in wheat, *Can. J. Plant Sci.,* 51, 187, 1971.

15. **Dickerson, G. E.,** Techniques for research in quantitative animal genetics, in *Techniques and Procedures in Animal Science Research,* American Society of Animal Science, Albany, N.Y., 1969, 36.

16. **Cockerham, C. C.,** Covariances of relatives from self-fertilization, *Crop Sci.,* 23, 1177, 1983.

17. **Hayes, J. F. and Hill, W. G.,** A reparameterization of a genetic selection index to locate its sampling properties, *Biometrics,* 36, 237, 1980.

18. **Robinson, H. F., Comstock, R. E., and Harvey, P. H.,** Genotypic and phenotypic correlations in corn and their implications in selection, *Agron. J.,* 43, 282, 1951.

19. **Yousaf, M.,** The use of selection indices in maize *(Zea mays),* in *Genetic Diversity in Plants,* Muhammed, A., Aksel, R., and von Borstel, R. C., Eds., Plenum Press, New York, 1977, 259.

20. **Searle, S. R.,** *Linear Models,* John Wiley & Sons, New York, 1971.

21. **Thompson, W. A., Jr.,** The problem of negative estimates of variance components, *Ann. Math. Stat.,* 33, 273, 1962.

22. **Corbeil, R. R. and Searle, S. R.,** A comparison of variance component estimators, *Biometrics,* 32, 779, 1976.

23. **Lee, K. R. and Kapadia, C. H.,** Variance component estimates for the balanced two-way mixed model, *Biometrics,* 40, 507, 1984.

Chapter 5

ISSUES IN THE USE OF SELECTION INDICES

I. INTRODUCTION

A common assumption in developing selection indices is that genotypic and phenotypic values are distributed with a multivariate normal distribution. If so, and if the parameters of that distribution are known, the procedures introduced in Chapter 1 will lead to maximum improvement in any specified linear function of genotypic values. In practice, characteristics will not likely be distributed exactly according to the normal distribution. The question of normality is considered in this chapter. This question is closely associated with the question of whether or not relationships between genotypic and phenotypic values are linear.

In general, index selection is restricted to worth functions which are linear. This restriction can be serious in improving some traits and is discussed briefly.

Estimates of population parameters are required in developing selection indices. Estimated indices will not give as great an overall response to selection as would an index based on true population parameters. It is important to know how large a sample is required to develop precise estimates of population parameters, as well as knowing how serious it is to proceed with estimates of questionable precision. These issues are also addressed in this chapter.

II. NORMALITY OF GENOTYPIC AND PHENOTYPIC DISTRIBUTIONS

In developing a selection index, it is assumed that genotypic and phenotypic values are normally distributed. Two questions can be asked. First, how critical is this assumption to the development and application of a selection index, and, second, do genotypic and phenotypic values show distributions which are sufficiently close to normal? Cochran[1] showed that, if there exists a regression of genotypic worth on phenotypic values, the optimum method for improving genotypic worth is to select on the basis of the linear regression. From Cochran's paper, the assumption of normality appears essential only for establishing a simple relationship between selection intensity and the standardized selection differential, and for predicting the impact of selection on subsequent changes in genotypic and phenotypic parameters.

Since the index is chosen so as to maximize the correlation between genotypic worth and the index, Harville[2] suggested that selection should be based on the index, even in the absence of normality. Again, the assumption of normality is not especially critical to the development and application of selection indices unless one wishes to predict changes in genotypic variances and covariances.

In practice, the genotypic effects of alleles at different loci are distributed according to a multinomial distribution. However, the genotypic value of an individual or a line will be the overall sum of the effects at the individual loci. The distribution of the overall genotypic values will approach a normal distribution if the number of loci is large. Figure 1 shows the cumulative distributions of genotypic values for 1, 3, and 5 independent loci with equal additive effects, two alleles per locus, and allele frequencies of 0.5 at all loci. With as few as five loci, the distribution of genotypic values is close to that of a normal population with the same mean and variance. Of course, the discrepancies would be greater than illustrated if the population were inbred, if the genotypic values were not equal at all loci, or if the allele frequencies were not 0.5. Nevertheless, the distribution of genotypic values can be expected to quickly approach a normal distribution as the number of genes contributing to the overall genotypic value increases.

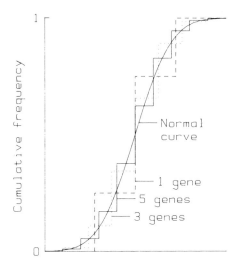

FIGURE 1. Cumulative frequency distributions of genotypic values for 1, 3, and 5 additive loci with equal effects and gene frequencies of 0.5 compared to a normal distribution having the same mean and variance.

Table 1
SKEWNESS AND KURTOSIS OF F_2 GENOTYPIC
VALUES IN RELATION TO NUMBER OF LOCI WITH
EQUAL EFFECTS AND ALLELE FREQUENCIES OF 0.5

	Genetic model			
	Equal additive		Equal dominant	
Number of loci	Skewness[a]	Kurtosis[a]	Skewness	Kurtosis
1	0.0	−1.00	−1.15	−0.67
2	0.0	−0.50	−0.82	−0.33
3	0.0	−0.33	−0.67	−0.22
4	0.0	−0.25	−0.58	−0.17
5	0.0	−0.20	−0.52	−0.13
6	0.0	−0.17	−0.47	−0.11
8	0.0	−0.12	−0.41	−0.08
10	0.0	−0.10	−0.37	−0.07
15	0.0	−0.07	−0.30	−0.04
20	0.0	−0.05	−0.26	−0.03

[a]   Coefficient of skewness = $m_3/(m_2)^{1.5}$ and coefficient of kurtosis = $m_4/(m_2)^2$, where m_2, m_3 and m_4 are the second, third, and fourth central moments of the distribution. Expected values for a normal distribution are 0.0 in each case.

Another way of examining the normality of the distribution of genotypic values is to compare the coefficients of skewness and kurtosis with those expected for a normal distribution. With independent loci carrying alleles which have equal additive effects, genotypic values in an F_2 population will have a coefficient of skewness of 0.0 and a coefficient of kurtosis which will approach the normal distribution value of 0.0 as the number of loci increases (Table 1). If plus (contributing to higher genotypic values) alleles are completely dominant over minus alleles, the distribution of genotypic values in an F_2 population will show a negative skewness (lower tail longer than upper tail) and negative kurtosis (flatter top than in a normal distribution) which will both tend toward the normal distribution values of 0.0 as the number of loci increases. The approach toward normality would not be so rapid if alleles at different loci had unequal effects or if they were linked.

As a guideline to interpreting Table 1, note that, for large samples from a normal distribution, the coefficient of skewness is approximately normally distributed with mean zero and variance 6/n, where n is the sample size. Similarly, the coefficient of kurtosis is normally distributed with mean zero and variance 24/n. Thus, in samples of n = 500 from a normal population, the coefficient of skewness would differ from its expected value of 0.0 by more than 0.074 (0.676 standard deviations) in half the samples. Similarly, the coefficient of kurtosis would be expected to deviate from its expected value of 0.0 by more than 0.148 in half the samples. In this context, the population distribution of F_2 genotypic values controlled by 15 or more loci would not deviate in kurtosis from the expected normal value any more than would a sample of 500 from a truly normal population. Of course, a finite sample of genotypic values from that population may deviate by a greater amount. If a trait is controlled by genes which show complete and unidirectional dominance, the distribution of genotypic values in the F_2 will differ markedly from that of a normal distribution in terms of skewness.

Genotypic values for most quantitative traits will have distributions that are sufficiently close to normal for methods developed on a normal assumption to be effective. Only in cases where dominant genotypic effects are important and unidirectional, and where expression of a trait is affected primarily by few genes with large effect, is it doubtful that the assumption can be accepted. Response to single trait selection under these conditions has been the subject of a series of papers by Latter.[3,4]

A second component of the overall assumption that phenotypic and genotypic values be normally distributed requires that environmental deviations be normally distributed. If normally distributed, the frequency of a particular value is inversely proportional to the square of its distance from the population mean. Many populations encountered in agricultural research have distributions which, to a good degree of approximation, follow this law. If phenotypic distributions show a normal distribution, one may presume that the environmental deviations are also normally distributed.

A second argument for accepting the assumption that environmental deviations are normally distributed is based on the central limit theorem of statistics. Briefly, this theorem states that, if a random variable, Y, is proportional to the sum of a large number of independent variables, the distribution of Y will approach that of a normal distribution. Thus, if many micro- and macro-environmental factors affect the phenotypic expression of a trait, one might reasonably assume that the net effects of all factors will be normally distributed.

Significant deviations from normality should only be expected when a trait is controlled by few loci, or at least is influenced primarily by a few loci with large effects. For most quantitative traits, it is doubtful that deviations from normality will be great. Even if the assumption of normality does not strictly apply to a number of traits in a population, it is likely that the strategies based on that assumption will still result in useful rules for multiple-trait selection in plant populations.

Perhaps the most critical part of the overall assumption of multinormality concerns the additivity of effects and the linearity of relationships among variables. The phenotypic value of a trait is assumed to equal the sum of the genotypic and environmental effects. Likewise, if two traits are related, it is assumed that the relationship between the two is linear.

There are many cases where the assumption of additivity is not strictly acceptable. If yields of a number of genotypes are more variable in high-producing environments than in low-producing environments, there is nonadditivity of the effects of genotype and environ-

Table 2
THE RELATIONSHIP BETWEEN THE PRODUCT OF TWO NORMAL VARIABLES AND THE LINEAR APPROXIMATION TO THAT PRODUCT

Component X		Component Y			Sum of squares (%) of
Mean	S.D.	Mean	S.D.	r_{XY}	Z = XY explained by Z'[a]
10	4	10	4	0.00	93.1
10	4	10	2	0.00	96.8
10	4	4	1	0.00	96.2
10	4	4	1.6	0.00	93.7
10	4	10	4	0.50	95.8
10	4	10	4	−0.50	87.0

[a] $Z' = \overline{X}Y + \overline{Y}X - \overline{XY}$.

ment. Yield may be the result of a multiplicative interaction between genotype and environment, and one should therefore expect nonlinearity in the relationship between phenotypic and genotypic values. Multiplicative interactions between component traits will also give nonlinearity. Since yield is the result of a multiplicative interaction between its components, the relationship between yield and any of its components cannot be strictly linear.

There may be instances where nonlinearity is present though not expected. In other cases, relationships which are believed to be nonlinear may actually be linear, or nearly so, because of the limited range of expression in the traits or because of correlations between the multiplicative components. The most likely type of nonlinearity is due to the multiplicative interaction of two or more traits. Thus, if $Z = XY$, there is a nonlinear relationship between Z and X, and between Z and Y. However, the degree of nonlinearity may not be great in practice. Consider that X and Y are normally distributed variables. If the values of X and Y are to be positive, the means for X and Y should be positive and their coefficients of variation should not exceed appproximately 43% (mean − 2.325 standard deviations > zero with probability = 0.99). Using these guidelines, a number of simulations were conducted in which 200 pairs of values (X,Y) were generated and Z was calculated as $Z = XY$. To assess the degree of nonlinearity in the relationship between Z and X, or Z and Y, the value of Z was compared to a linear approximation, Z' (based on a Taylor expansion about the mean values), where $Z' = \overline{X}Y + \overline{Y}X - \overline{XY}$. The results are summarized in Table 2.

Even in cases of high coefficients of variation and negative correlations between component traits, most of the variation in the product can be explained by a linear approximation. Even where expected, nonlinearity may not be great enough to seriously reduce the efficiency of techniques based on the assumption of linearity. Furthermore, it will be difficult to test for and detect cases of nonlinearity. In Table 2, it is shown that with equal variances of 4.0 and means of 10.0 for two independent component traits, 6.9% (100 − 93.1) of the variation in Z is related to nonlinearity of the relationship. If one were to compare Z to either of its components, 46.55% of the variation in Z would be linearly related to that component. Of the remaining 53.45% only 12.3% is due to nonlinearity; the remainder being due to the linear relationship with the other component. Under these circumstances, it would be extremely difficult to detect any degree of nonlinearity between the product and either of its components.

If variables have phenotypic distributions which are symmetric about the mean and continuous, it is doubtful that nonlinearity or nonnormality will be detectable in most experiments. Researchers should be aware of these assumptions and should avoid relying on methods based on normal distributions if the data are obviously nonnormal. In detectable cases of nonnormality or nonlinearity, it may be possible to transform the data to more closely approximate a normal curve. It is not possible to recommend an objective rule or routine procedure to check for these departures from the assumption of multinormality.

III. NONLINEAR WORTH FUNCTIONS

In most applications of selection indices, it has been assumed that, for a particular trait, genotypic worth either increases or decreases linearly with changes in genotypic value. Kempthorne and Nordskog[5] did allude to the possibility that genotypic worth might be a quadratic function of genotypic value. For example, $W = k(G - d)^2$ may be used to represent the relationship between genotypic worth, W, and the genotypic value, G, of a particular trait. They noted that this worth function could be rewritten as $W = kG^2 - 2kdG + kd^2$ and one could handle this in the normal index methodology by including squared values for the trait as a separate trait in all calculations.

Baker[6] described a problem in which the genotypic worth of a trait is zero unless the genotypic value exceeds some minimum value. Beyond that point, genotypic worth is a linear function of genotypic value. He recommended that a standard selection index be estimated for this situation, but, prior to calculating index scores, that phenotypic values for that trait be transformed by setting values below a certain critical level to zero. The choice of critical level for the transformation would depend upon the critical level in the worth function as well as heritability of the trait.

Yamada et al.[7] suggested that the proportional constraints method of Tallis[8] could be used instead of the quadratic approach suggested by Kempthorne and Nordskog.[5] With proportional constraints, the index is based on specified desired changes in each trait. If the population mean is less than the optimum genotypic value for a particular trait, the desired change for that trait would be specified as a positive value. On the other hand, a trait whose mean value exceeded the optimum level would receive a negative value for desired change. While it is true that this approach might have similar results to that of Kempthorne and Nordskog,[5] the two curves relating genotypic worth to genotypic value would differ somewhat. For Kempthorne and Nordskog's method, the curve is a smooth quadratic curve that may approach the optimum value at varying rates depending upon the parameter k. With proportional constraints, on the other hand, the curve is pointed in shape and may not be particularly well defined at its peak. In any case, there is no really satisfactory way of addressing the problem of nonlinear worth functions. The remainder of this book will deal only with linear functions of genotypic values.

IV. SAMPLE SIZES REQUIRED FOR ESTIMATING POPULATION PARAMETERS

Selection indices are based on estimates of population variances and covariances. Because variances and covariances are second order statistics, they are subject to large sampling errors. One should expect that large samples will be required to obtain acceptable degrees of precision. It is difficult to specify how large a sample must be to provide a sufficiently accurate estimate. One must be concerned with how well the sample estimates agree with the population values, as well as with how well they need to agree in order that the resulting indices be useful.

A sample variance, s^2, is an unbiased estimator of the corresponding population variance, σ^2. If the population has a normal distribution, the distribution of variances of samples of size n from the population can be related to the population variance by use of the chi-squared distribution. In fact, $(n - 1)s^2/\sigma^2$ is distributed as a chi-squared with $(n - 1)$ degrees of

Table 3
LOWER AND UPPER 95% CONFIDENCE LIMITS FOR THE RATIO OF
SAMPLE VARIANCE TO POPULATION VARIANCE IN RELATION TO
SAMPLE SIZE

Sample size (n)	$\chi^2_{0.975,n-1}/(n-1)$	$\chi^2_{0.025,n-1}/(n-1)$
10	0.30	2.11
20	0.47	1.73
30	0.55	1.58
50	0.65	1.42
100	0.74	1.30
200	0.81	1.20
400	0.86	1.14
1000	0.91	1.09

Table 4
LOWER AND UPPER 95% CONFIDENCE LIMITS FOR
THE POPULATION CORRELATION COEFFICIENT FOR
DIFFERENT SAMPLE CORRELATIONS AND SAMPLE SIZES

	Sample correlation					
	0.0		0.4		0.8	
Sample size	Lower[a]	Upper[a]	Lower	Upper	Lower	Upper
10	−0.57	0.57	−0.26	0.93	0.33	0.93
20	−0.42	0.42	−0.05	0.69	0.54	0.90
30	−0.35	0.35	0.05	0.65	0.60	0.89
50	−0.27	0.27	0.13	0.60	0.66	0.88
100	−0.19	0.19	0.22	0.55	0.71	0.86
200	−0.14	0.14	0.28	0.51	0.74	0.84
400	−0.10	0.10	0.31	0.48	0.76	0.83
1000	−0.06	0.06	0.35	0.45	0.78	0.82

[a]   Based on a modified z-transformation from Hotelling.[9]

freedom. This relationship can be used to construct a 95% confidence interval on the ratio of the sample variance to the true population variance for a given sample size; i.e.,

$$\text{Prob}\left[\frac{1}{n-1}\chi^2_{0.975,n-1} < \frac{s^2}{\sigma^2} < \frac{1}{n-1}\chi^2_{0.025,n-1}\right] = 0.95$$

Thus, if one wishes to choose a sample size which will estimate variances within, say, 10% of the population variance, n must be such that $\chi^2_{0.975,n-1}/(n-1) > 0.90$ and $\chi^2_{0.025,n-1}/(n-1) < 1.10$. One would require a sample size of something over 400 to be 95% confident that the sample variance would be within 10% of the true population variance (Table 3).

While it is not possible to use a similar approach to determine the sample size needed for estimating covariances between variables, some insight may be provided by considering confidence intervals based on sample correlation coefficients. The lower and upper 95% confidence limits[9] for the population correlation coefficient are tabulated for various sample correlation coefficients and sample sizes in Table 4. A sample size of several hundred is required to be confident that the population correlation coefficient deviates from the observed sample correlation by no more than 0.10.

Table 5
**EFFECT OF SAMPLE SIZE AND NUMBER OF
INDEPENDENT VARIABLES ON THE 2.5 AND
97.5 PERCENTILES OF THE DISTRIBUTION
OF THE SAMPLE MULTIPLE CORRELATION
COEFFICIENT WHEN THE POPULATION
CORRELATION IS ZERO**

| | Number of independent variables | | | | | |
| | 1 | | 3 | | 5 | |
Sample size	2.5	97.5	2.5	97.5	2.5	97.5
20	0.00	0.50	0.14	0.65	0.27	0.75
50	0.00	0.32	0.09	0.43	0.16	0.49
100	0.00	0.22	0.06	0.30	0.09	0.36
200	0.00	0.16	0.04	0.22	0.06	0.25

In most practical applications, several variances and covariances must be estimated simultaneously. One would expect the overall level of precision to decrease as the number of variables increases. One way to assess the impact of increasing the number of variables is to consider the distribution of the sample multiple correlation coefficient. When the population multiple correlation is zero, the sample multiple correlation coefficient (R) is related to the F-ratio for the mean square due to regression on k independent variables tested against the residual mean square with $(n - k - 1)$ degrees of freedom.[10] In fact, $[(n - k - 1)R^2]/[k(1 - R^2)] = F_{n, n-k-1}$. Use of this relationship, along with tables of the F-distribution, provides a basis for relating the lower and upper critical values of the distribution of the sample correlation coefficient to both sample size and the number of independent variables (Table 5). As indicated in Table 5, the range for the sample multiple correlation coefficient will become greater as the number of independent variables increases. Table 5 also shows that sample correlation coefficients give biased estimates of the population correlation coefficient, and that this bias increases as the number of independent variables increases. For a sample size of 100 and a population multiple correlation of 0.0, the 97.5 percentile of the distribution of the sample correlation coefficient is 0.22 with one independent variable, and 0.36 when five are included. This increase is indicative of the overall decrease in precision as the number of variables included in a selection index increases.

Genotypic variances and covariances required in index selection are estimated by linear functions of two or more variances and covariances. They will be even more sensitive to sample sizes than indicated by the above discussion. Hill and Thompson[11] discussed one particular aspect of this problem. If estimates of genotypic variances and covariances are sufficiently imprecise, ordinary or partial genotypic correlations may exceed the bounds of -1 to $+1$. That this problem exists in a particular matrix of genotypic variances and covariances is indicated if the matrix is not "positive definite".

Hill and Thompson[11] considered the case where s genotypic groups are tested in a one-way design with observations being made on p traits on each of n individuals within each group. By using both theoretical and computer simulation methods, they were able to show that there was a high probability that the genotypic variance-covariance matrix would not be positive definite even with quite large sample sizes. For example, if the intraclass correlations (heritabilities) were equal to 0.125 for all traits, if all traits were genotypically and phenotypically uncorrelated, and if 20 individuals were evaluated within each of 20 genotypic groups, they estimated that the probability of a nonpositive definite genotypic matrix was 38.2% if 6 traits were measured. That is, for experiments of this size, 38.2% of such experiments would be expected to give one or more estimates of heritability outside the range of 0 to 1, or of genotypic or partial correlations outside the range of -1 to $+1$. Doubling the number of observations within each genotypic group would reduce the probability of a nonpositive definite matrix to 4.4%. Doubling the number of genotypic groups to 40 and maintaining the number of individuals within each group at 20 would reduce the probability to nearly zero. Hill and Thompson's results suggest that large sample sizes are required, especially in terms of numbers of genotypic groups evaluated.

In further developing an approach to the problem of errors in index selection, Hayes and Hill[12] proposed a method by which phenotypic and genotypic covariance matrices could be transformed into a simple form. On the transformed scale, heritabilities must fall in the range of 0 to 1. Furthermore, comparing roots of the transformed genotypic covariance matrix with relative economic weights might serve to identify cases where the estimated index would be particularly unreliable.

This research led Hayes and Hill[12] to propose that estimated genotypic and phenotypic covariance matrices be adjusted to minimize problems caused by sampling errors. Their method was referred to as "bending" and consists of adjusting extreme values toward their mean. Both approximate theory and Monte Carlo simulation indicated that bending should result in the development of more efficient indices. The method may not always be easy to apply in practice. There are no accepted rules for deciding on the best bending factor. In cases where some estimates are outside their valid limits, one should select a bending factor which will at least bring these estimates into an acceptable range.

Nanda[13] first discussed whether or not the use of estimated population parameters for constructing selection indices would limit their efficiency. This problem is critical, and has since been evaluated by Cochran,[1] Williams,[14,15] Harris,[16,17] Hill and Thompson,[11] and Hayes and Hill.[12]

In evaluating the application of selection indices, three different types of selection responses must be recognized. First is the response that would occur if the index based on true population parameters (i.e., the optimum index) could be used. The second type of response is that which would occur if the index based on sample estimates (i.e., the estimated index) were used. And finally, there is the estimated response to selection based on the estimated index. Generally, actual responses to estimated indices will always be less than response to the optimum index. In cases where particularly poor estimates are available, actual response to an estimated index may even be in the wrong direction.[14] Improved estimates of population parameters usually require evaluation of a large number of genotypes.

Cochran[1] addressed the question of sample size by comparing the average correlation between genotypic worth and indices estimated from samples of various size to the correlation that would be expected if the true population parameters were used for constructing the index. He showed that, for a population with a multivariate normal distribution of genotypic and phenotypic values, the average sample correlation was equal to the population correlation multiplied by a factor that could never be greater than 1.0. This factor represents the average proportion of maximum response to index selection that can be achieved using indices based on estimates from samples. This factor will decrease as sample size decreases, and as the number of traits in the selection index increases. In an example, Cochran considered the question of how large a sample would be required to assure that the average sample correlation would not deviate from the population correlation (assumed to be equal to 0.7 or larger) by more than 5%. His calculation showed that the index coefficients should be estimated from a sample size of at least 45 genotypic groups. While the ratio is independent of the number of observations on each genotype, total response (even if the population parameters are known) will depend on both the number of observations per genotypic group as well as the number of genotypic groups.

Cochran's[1] formula was used to develop a table of the number of genotypic groups that

Table 6

NUMBER OF GENOTYPIC GROUPS THAT MUST BE SAMPLED IN CONSTRUCTING A SELECTION INDEX SO THAT AVERAGE RESPONSE WILL BE AT LEAST 90% OF OPTIMUM

Population correlation[a]	Number of traits in the index				
	2	3	4	5	6
0.8	7	10	13	16	19
0.7	10	15	19	24	30
0.6	14	22	28	37	45
0.5	21	34	44	58	71
0.4	35	57	73	96	119
0.3	64	106	135	179	223

[a] Correlation between genotypic worth and the optimum index.
Calculations based on formulae given by Cochran.[1]

should be evaluated in order to attain, on the average, a response of not less than 90% of that expected if the population parameters were known (Table 6). The data show that larger sample sizes are required if more traits are included in an index. Moreover, sampling of fewer than 20 to 30 genotypes will not give a response which approaches the maximum except when few traits are included in the index and when the population correlation coefficient between genotypic worth and the optimum index is high.

As indicated above, Cochran[1] concluded that the correlation between genotypic worth and an estimated index would be equal to the correlation between genotypic worth and an optimum index multipled by a factor which can never exceed 1.0. This factor was shown to equal $t[t^2 + R - 1]^{0.5}$, where R is the correlation between genotypic worth and the optimum index based on true population parameters, and t follows a noncentral t-distribution. The noncentral t-distribution has $(p - 1)$ degrees of freedom and noncentrality parameter $R[n - p - 1]^{0.5}/[1 - R^2]^{0.5}$, where p equals the number of traits included in the index. In order to illustrate the proportion of maximum response that can be expected from estimated indices, the methods outlined by Cochran[1] were used to calculate the ratio of the correlation between genotypic worth and the estimated index to the correlation between genotypic worth and the optimum index (Table 7).

The data in Table 7 show the effects of sample size used to estimate population parameters, and number of traits included in the index, on the expected efficiency of the estimated index. For example, if population parameters are estimated from a sample of 40 genotypes from a specified population, and if the correlation between genotypic worth and an index based on known population parameters is 0.5, an estimated selection index for four traits would have a relative efficiency of 0.88. That is, the estimated index would be only 88% as efficient as would the optimum index, if it were known. It can be seen that an index based on a small sample size, say 20 genotypes, and including several traits, may be substantially less efficient than desired, particularly if the correlation between genotypic worth and the optimum index is not high.

It should be noted that the formulation given by Cochran[1] does not include any direct accounting of extent of replication of each genotype. This item is reflected in the magnitude of the correlation between the genotypic worth and the optimum index. If replication is limited, this correlation will be small and it will be even more important to estimate population parameters from sufficiently large samples.

Table 7
RATIOS OF CORRELATION BETWEEN GENOTYPIC WORTH AND AN ESTIMATED INDEX TO CORRELATION BETWEEN GENOTYPIC WORTH AND AN OPTIMUM INDEX AS INFLUENCED BY NUMBER OF TRAITS AND SAMPLE SIZE

	Number of traits in the estimated index			
Sample size	2	4	6	8
	$(R^a = 0.5)$			
20	0.89	0.76	0.65	0.56
40	0.96	0.88	0.82	0.76
60	0.97	0.92	0.88	0.83
	$(R^a = 0.8)$			
20	0.98	0.95	0.90	0.85
40	0.99	0.98	0.96	0.94
60	0.99	0.99	0.97	0.96

[a] R = correlation between genotypic worth and optimum index.

Williams[14,15] also investigated the problem of using estimated parameters for constructing selection indices. Williams compared expected response from the use of an estimated index with that expected from the use of a base index in which relative economic values were used as index coefficients. By comparing expected correlations of the two types of indices with the optimum index, Williams[14] concluded that estimates of genotypic and phenotypic parameters must be quite good for the estimated index to be more effective than the base index.

Williams[15] showed that, for the two-variate case, the correlation between the estimated index and genotypic worth would actually be negative if sample sizes used in developing the index were too small. For two traits, if the number of genotypes used to develop the estimated index is less than $(3 - R^2)/R^2$ where R is the correlation between genotypic worth and the optimum index, the estimated index will on average be worse than random selection of genotypes.

Williams[15] also observed that coefficients of the estimated index are biased estimates of the optimum coefficients. It was suggested that the bias could be removed by applying an adjustment factor based on the number of genotypes tested.

Harris[16,17] used computer simulation to assess the consequences of sampling errors on response to index selection. Simulation results for various combinations of parameters for two traits showed that responses to estimated selection indices are expected to be extremely unreliable when heritabilities are low, when correlations between traits are low or negative, and when estimates are based on small samples of genotypes. Harris[16] developed approximate formulas for the mean and variance of expected response to selection with estimated indices.

V. SUMMARY

The assumption of multinormality does not seem overly critical in the development and application of selection indices. If traits are controlled by a number of genes and are subject to a moderate amount of environmental variation, both genotypic and phenotypic distributions should be sufficiently close to normal for the selection rules based on normal theory to be useful in crop inprovement programs.

Nonlinear worth functions have been considered. While such definitions of genotypic worth may be valuable in crop improvement, they are difficult to deal with in practice. At best, ad hoc methods may allow some approximation to an optimum strategy when genotypic values have optima or upper or lower critical bounds.

It has been well established that index coefficients based on estimated variances and covariances will give less overall response in genotypic worth than would an optimum index. If very poor estimates are used, response may be negligible or even in the wrong direction. Review of papers dealing with the use of sample estimates for selection indices suggests that estimates must be based on a minimum sample of 30 to 40 genotypes. Multivariate tests have been developed for evaluating the utility of estimated variance-covariance matrices and may serve to avoid the use of grossly inadequate indices.

REFERENCES

1. **Cochran, W. G.,** Improvement by means of selection, in Proc. 2nd Berkeley Symp. Math. Stat. Prob., 1951, 449.
2. **Harville, D. A.,** Optimal procedures for some constrained selection problems, *J. Am. Stat. Assoc.,* 69, 446, 1974.
3. **Latter, B. D. H.,** The response to artificial selection due to autosomal genes of large effect. I. Changes in gene frequency at an additive locus, *Aust. J. Biol. Sci.,* 18, 585, 1965.
4. **Latter, B. D. H.,** The response to artificial selection due to autosomal genes of large effect. II. The effects of linkage on limits to selection in finite populations, *Aust. J. Biol. Sci.,* 18, 1009, 1965.
5. **Kempthorne, O. and Nordskog, A. W.,** Restricted selection indices, *Biometrics,* 15, 10, 1959.
6. **Baker, R. J.,** A simple method for incorporating non-linear worth functions into index selection, *Agron. Abstr.,* 1970, 5.
7. **Yamada, Y., Yokouchi, K., and Nishida, A.,** Selection index when genetic gains in individual traits are of primary concern, *Jpn. J. Genet.,* 50, 33, 1975.
8. **Tallis, G. M.,** A selection index for optimum genotype, *Biometrics,* 18, 120, 1962.
9. **Hotelling, H.,** New light on the correlation coefficient and its transforms, *J. Royal Stat. Soc.,* Sec. B, 15, 193, 1953.
10. **Morrison, D. F.,** *Applied Linear Statistical Methods,* Prentice-Hall, Englewood Cliffs, N.J., 1983.
11. **Hill, W. G. and Thompson, R.,** Probabilities of non-positive definite between-group or genetic covariance matrices, *Biometrics,* 34, 429, 1978.
12. **Hayes, J. F. and Hill, W. G.,** Modification of estimates of parameters in the construction of genetic selection indices ("Bending"), *Biometrics,* 37, 483, 1981.
13. **Nanda, D. N.,** The standard errors of discriminant function coefficients in plant breeding experiments, *J. Royal Stat. Soc.,* Sec. B, 11, 283, 1949.
14. **Williams, J. S.,** The evaluation of a selection index, *Biometrics,* 18, 375, 1962.
15. **Williams, J. S.,** Some statistical properties of a genetic selection index, *Biometrika,* 49, 325, 1962.
16. **Harris, D. L.,** The influence of errors of parameter estimation upon index selection, in Statistical Genetics and Plant Breeding, Publ. No. 982, Hanson, W. D. and Robinson, H. F., Eds., National Academy of Sciences, National Research Council, Washington, D.C., 1963, 491.
17. **Harris, D. L.,** Expected and predicted progress from index selection involving estimates of population parameters, *Biometrics,* 20, 46, 1964.

Chapter 6

USE OF SELECTION INDICES AND RELATED TECHNIQUES FOR EVALUATING PARENTAL MATERIAL

I. INTRODUCTION

Conventional index methodology is based on a multivariate statistical technique related to discriminant analysis. Related, but distinct, techniques have been recommended for choosing parental material for crop improvement programs. The related techniques include vector analysis,[1] canonical analysis,[2] and multivariate distance analysis.[3]

The first two techniques are based on the hypothesis that most quantitative traits are controlled largely by genes with additive effects. Thus for each trait the mean for a population generated in a plant breeding program should be close to the mean of the parents used to produce that population. Vector analysis and canonical analysis provide objective approaches to predicting population means. Those who favor these methods imply that the best combination of parental material will be one which produces a population whose mean for each trait is closest to the objective of the breeding program.

The generalized distance (D^2) method differs in that its use is restricted to choosing parental lines which are diverse in a multivariate sense. Use of diverse parents will give populations with greater genetic variation and may increase the prospect for isolating lines with superior attributes.

The different emphasis of the generalized distance approach will identify and group different sets of parents than would vector or canonical analysis. Furthermore, choosing parents so as to optimize population means may be the best strategy for short-term (one or few cycles) breeding programs, while choosing parents with maximum genetic diversity may be the best strategy for long-term recurrent selection programs. This argument is based on the observation that populations produced from crosses between adapted and unadapted material, while having greater genotypic variation, often have means which are inferior to populations generated by crossing adapted parents. This, coupled with the fact that only a portion of the genotypic variation is captured in improved performance in short-term selection programs, explains the frequent observation that the best genotypes are isolated from crosses among highly adapted parents.

A hypothetical example might serve to clarify the argument. Considering a single quantitative trait, let the genotypic values of inbred parents A, B, and C be 10, 20, and 24, respectively. Assume that the average genotypic values of inbred lines developed after crossing each pair of parents is equal to the mid-parent value; i.e., 15 for A × B, 17 for A × C, and 22 for B × C. Assume further that the genotypic variances among lines within crosses are 5, 7, and 2 for crosses A × B, A × C, and B × C, respectively, and that short-term selection is expected to give a response equal to 0.5 genotypic standard deviations. With this scenario, one would expect the average genotypic values of lines selected within cross A × B to be equal to $15 + 0.5(5)^{0.5} = 16.1$, with $17 + 0.5(7)^{0.5} = 18.3$ and $22 + 0.5(2)^{0.5} = 22.7$ being the corresponding values for crosses A × C and B × C. In this example, the cross with the least genotypic variation gives the best selected lines by virtue of its higher cross mean. In fact, one would have to obtain responses of over eight genotypic standard deviations before the wide cross (A × C) would yield lines superior to those of B × C. Similar expectations apply to instances where response is measured in terms of a linear function of several traits.

The method used for evaluating parental material may well depend on whether or not one contemplates a short-term or a long-term breeding program. With short-term breeding programs, population means are critical. In these cases, the techniques of vector analysis and canonical analysis should be considered. If response to recurrent selection in a long-term breeding program is expected to be sufficient to make within-population genotypic variability relatively more important than the population mean in determining final outcome, the generalized distance method should be considered.

All of these methods rely to some degree on the assumption of additivity of gene effects. Vector analysis and canonical analysis rely on this assumption to the extent that the population mean be equal to the mean of its parents. If the population consists of inbred lines developed by crossing two or more inbred parents, this assumption is equivalent to assuming that additive types of epistasis are unimportant in the inheritance of quantitative traits. Grafius[1] argued that this type of epistasis usually indicates that a trait is a multiplicative construct of two or more component traits. He cited the case of cereal yields, where yield is the product of number of heads per unit area, number of kernels per head, and weight per kernel. In such a case, the assumption concerning absence of epistasis can often be circumvented by considering the components themselves, rather than yield.

The generalized distance method requires that the generalized distance between parents reflect the genotypic variance within the population developed by crossing those parents. While it may not be strictly necessary to assume additive gene action in this case, certain types of epistasis will reduce the relationship between parental range and population variance.

II. METHODS FOR CHOOSING PARENTS TO PRODUCE POPULATIONS WITH DESIRABLE MEANS

A. Vector Analysis

Grafius[1] gave a detailed description for using vector analysis to choose parents for a plant-breeding program. The method has not become popular with plant breeders because many plant breeders are not comfortable with the geometric approach used by Grafius.

The method proposed by Grafius[1] includes several important elements. The first concerns the relative weights that are to be assigned to traits in the breeding program. In using the conventional selection index, these weights are known as relative economic weights, and are developed by a somewhat subjective assessment of the economic value of each trait. Grafius proposed a novel method for developing relative weights. He recommended that a representative sample of inbred lines or other parental material be evaluated for all the characteristics which are considered to be important in the breeding program. In addition, one or more experienced plant breeders should be asked to assign an overall worth score to each line. This might be done by comparing each trait to a subjective yardstick. The relative weights for each trait would then be determined by calculating the multiple regression of the overall worth score on observed phenotypic values of all traits. This exercise serves to quantify the unstated subjective weights that knowledgeable plant breeders use in evaluating parental material. The standardized partial regression coefficients from the regression analysis would then be used as estimates of the relative weights to be attached to changes of one standard deviation in each of the traits.

Grafius[1] did discuss the impact of genotype-environment interaction and environmental variability on the estimated weights. He suggested that estimated weights should be adjusted if the correlations between environments or between seasons differed greatly from one trait to another.

A second element of Grafius' proposal[1] concerns transformation of data. For each trait, the data are transformed so that the mean value is 1.0, and so that those traits with greater relative importance show greater variation about 1.0 than do those with less relative importance.

In evaluating a set of potential parents, data are collected for all traits in each potential parent. Moreover, the breeder must specify an "ideal" value for each trait. The data from each of the potential parents, and the specified values for one or more ideal genotypes, are then transformed to meet the requirement stated above. Correlations (of transformed values for all traits) are then calculated for all possible pairs of potential parents and ideals. At this point, Grafius used vector analysis to determine which parents would produce a population whose means would most closely approach those of the ideal genotype. In essence, the resultant of the two vectors representing two potential parents is compared to the vector representing the ideal genotype. This comparison gives an indication of how closely the resulting population would approach the ideal, as well as the relative proportions of the parents that should be used in the cross.

In a more familiar context, the procedure is equivalent to using transformed data to calculate the multiple regression of the ideal on the two parents being considered. The coefficient of determination will indicate how close the population mean will be to the ideal while the partial regression coefficients will indicate the relative proportions of each parent which should be used in creating the population. With this type of presentation, it is quite simple to extend the basic concept to a consideration of multiple parent crosses.

In a worked example, Grafius[1] considered 15 characteristics measured in each of 31 barley, *Hordeum vulgare*, lines. The correlation (of transformed values for 15 characteristics) between potential parents A and C was 0.21, while the correlation between A and a specified ideal was 0.86, and between C and the ideal was 0.33. A consideration of a vector diagram as well as sines, cosines, and tangents of various angles, led Grafius to conclude that a population developed by having proportions 0.84 of parent A and 0.16 of parent C would give a population whose transformed means would show a correlation of 0.88 with the ideal.

The same conclusion can be derived from multiple regression. The relative proportions p_A and p_C of the two parents are the standardized partial regression coefficients which can be estimated from the following two simultaneous equations.

$$p_A + 0.21 \ p_C = 0.86$$

$$0.21 \ p_A + p_C = 0.33$$

Solving these two equations gives $p_A = 0.84$ and $p_C = 0.16$. Moreover, the coefficient of determination, calculated as

$$(0.84)(0.86) + (0.16)(0.33) = 0.77$$

gives the square of the correlation between the ideal and the weighted average of the two parents ($0.77 = 0.88^2$).

Grafius[1] pointed out that, in practice, one would use two backcrosses to parent A to produce a population with relative proportions of 0.875 of parent A and 0.125 of parent C. The correlation with the ideal would then be

$$[(0.875)(0.86) + (0.125)(0.33)]^{0.5} = 0.88$$

within rounding error of the optimum value.

The vector method for choosing parents includes assigning weights to traits to reflect their relative importance as indicated by a knowledgeable plant breeder, as well as their relationships with other traits and their constancy over environments. The method also requires that the plant breeder specify goals for the breeding program in terms of an ideal genotype. The method provides an indication of which parents are expected to produce a population which most closely approximates the specified ideal and provides guidelines as to how the parents should be crossed. The method described by Grafius[1] is not easily applied to choosing more than two parents per cross, but can be modified to do so.

As presented, vector analysis is most appropriate for choosing parents in breeding programs for inbred species where the new cultivar is to be an inbred line. Grafius[1] did suggest that the method would be useful in a program for developing hybrids. If heterosis is due primarily to multiplicative interaction of additive component traits, vector analysis based on the component traits could be used to predict hybrid performance. If the value of the hybrid depends primarily on the recombining of several additive characteristics into one homogeneous genotype, vector analysis can provide a method for assessing all characteristics simultaneously. Furthermore, Grafius pointed out that vector analysis could be used in cross-fertilized species by using topcross or polycross progeny data as a basis for parental evaluation.

B. Least-Squares Analysis

Pederson[4] pointed out that a least-squares approach can be used as an alternative to the vector method. As implied in the above section, it is possible to translate the vector method[1] into terminology dealing not with geometry of vectors, but with the algebra of multiple regression. The proposal by Pederson[4] is similar to a multiple regression approach.

Pederson[4] also introduced a different method for specifying the relative importance of different traits. Besides specifying an ideal value for each trait, Pederson recommended that one also specify a ''preferred maximum deviation''. The squares of these deviations would then be used as weights in a weighted least-squares procedure to estimate the proportion of each of two or more parents that would be required to produce a population whose means most closely approximate those of the ideal. In using this technique, one would develop least-squares estimates for all possible combinations of parents, and choose that set with the lowest weighted sum of squares of deviations between the predicted population means and the ideal. As with the vector method, the least-squares method presumed that the population means are equal to the means of the parents, each weighted by their relative contributions to the populations.

Pederson[4] also suggested an enumeration scheme for summarizing the least-squares evaluation of potential parents. A series of $<$ and $>$ symbols can be used to indicate whether the expected population mean is less than or greater than the ideal value by one or more multiples of the preferred maximum deviation for each trait. In cases where two or more crosses are expected to produce populations with means close to the ideal, Pederson emphasized the importance of choosing the cross or crosses whose parents show the greatest divergence for those traits for which directional selection is to be practiced, and least divergence for those traits for which little selection pressure is to be exerted.

C. Canonical Analysis

Canonical analysis is a multivariate statistical method for representing data in dimensions which give the greatest discrimination among groups of individuals. The basic data are transformed so that the first canonical variate (a linear function of the original measurements) shows the greatest ratio of among-group to within-group variation. The second canonical variate shows the second greatest possible ratio, and so on. Each succeeding canonical variate is chosen so as to be uncorrelated over group means with all preceeding canonical variates.

This procedure can best be described by reference to simulated data (Table 1). In this example, two traits have been measured on each of ten potential parents, each tested in two replications of a randomized complete block design. For the raw data, analysis of variance showed that the ratio of the between-parent mean square to the error mean square was 3.07 for the first trait and 1.44 for the second. For these two traits, both the between-parent and error terms were correlated (0.51 and 0.80, respectively). When canonical analysis is applied to the data in Table 1, the data are transformed to two new variables, X and Y, by the following equations.

Table 1
SIMULATED DATA FOR TWO
TRAITS MEASURED ON TEN
PARENTS EVALUATED IN TWO
REPLICATIONS

Parent	Replication	Trait A	Trait B
1	1	89.5	999.5
1	2	88.7	1002.4
2	1	102.8	983.1
2	2	98.4	975.5
3	1	111.8	1048.0
3	2	102.7	1002.2
4	1	93.2	1004.2
4	2	103.9	1029.1
5	1	104.7	1033.2
5	2	118.4	1044.0
6	1	104.1	1038.4
6	2	100.3	1009.0
7	1	102.3	1027.5
7	2	102.1	1002.4
8	1	95.5	1047.2
8	2	87.9	980.4
9	1	101.5	1013.9
9	2	100.2	1003.8
10	1	96.1	1033.2
10	2	101.3	1019.1

$$X = 0.975 \text{ A} - 0.222 \text{ B}$$

$$Y = 0.193 \text{ A} + 0.981 \text{ B}$$

Analysis of the transformed variables, X and Y, showed that the ratio of the between-parents mean square to the error mean square was 6.39 for X and 1.43 for Y. Moreover, the mean cross-products for both between-parents and error were near zero. Clearly, the transformation creates a new variable, X, which accounts for a major portion of the total variation among the parents. Rather than having to consider two traits, A and B, it may suffice to be concerned with the one new variable.

Whitehouse[5] noted that a large portion of the variation of six traits measured on 36 genotypes of barley could be summarized by two or three canonical variates. He proceeded to plot the values of the first two canonical variates for each parent and each of their crosses in order to obtain a general impression of the relationships between parents and their progeny.

Whitehouse[5] described the use of canonical analysis for choosing parents for a breeding program. The objective of the method is to predict parental combinations which will give progeny populations with means as close as possible to those of a specified ideal breeding target. Canonical analysis is used to reduce the dimensionality of the data to perhaps as few as two canonical variates. The relationships among the parents can then be visualized with a plot of the first two canonical variates. Assuming that a population derived by crossing two parents will lie midway between them on the canonical diagram, one can then compare the predicted population mean to the ideal.

The method of canonical analysis is similar to that of vector analysis, or least-squares analysis, in that several traits are considered simultaneously, the population mean is assumed to equal the mid-parent value, and parents are chosen by comparing the expected population mean to an ideal. The method differs, however, in that the plant breeder has no control over the relative weighting of the various traits. In canonical analysis, traits are weighted according to how well they discriminate between parental genotypes. This may or may not be related to the relative importance of traits as perceived by the plant breeder.

Riggs[6] gave a more rigorous description of canonical analysis and applied it to selection among 100 lines from a composite population of barley. He found that five canonical variates were required to describe 96% of the total variation observed in seven traits measured in a replicated trial of 120 lines and six standard cultivars. Canonical variates derived from the replicated experiments were then calculated for each of 900 unreplicated lines. Lines whose canonical variates were within the range of those of the standard cultivars were selected for further evaluation.

Riggs[6] cautioned that the analysis may not reflect the relative worth of the traits measured. He recommended that the breeder avoid measuring traits which give obvious but unimportant differences among groups.

III. METHODS FOR CHOOSING PARENTS HAVING GREATER GENETIC DIVERSITY

Bhatt[3,7] used generalized distance (D^2) measurements to try to identify genetically diverse parents among a group of 40 wheat, *Triticum aestivum*, genotypes. The method made use of replicated measurements on each genotype and assumed that the covariances among traits were the same for all genotypes. D^2_{ij} estimates were used to measure the multi-dimensional distance between parents i and j, standardized relative to the variation within parental genotypes.

The data in Table 1 can be used to demonstrate the essential elements of multivariate distance analysis. From a randomized block analysis of covariance of that data, the error mean squares and cross-products (with 9 degrees of freedom [d.f.]) are, in matrix form,

$$W = \begin{bmatrix} 28.1 & 81.4 \\ 81.4 & 366.0 \end{bmatrix}$$

with inverse

$$W^{-1} = \begin{bmatrix} 0.1002 & -0.0223 \\ -0.0223 & 0.0077 \end{bmatrix}$$

Based on the inverse of the error matrix, the formula for the distance between two parents, the ith and the jth, is $D^2_{ij} = 0.1002(A_i - A_j)^2 + 2(-0.0223)(A_i - A_j)(B_i - B_j) + 0.0077(B_i - B_j)^2$. Thus, for parent 2 (Table 1) the mean for trait A is 100.6 and the mean for trait B is 979.3. For parent 8, the mean values for traits A and B are 91.7 and 1013.8, respectively. Inserting these values into the distance equation gives $D^2_{28} = 30.8$, the greatest distance for any pair of the 10 potential parents. By contrast, the multivariate distance between parent 7 (means = 102.2 and 1014.95) and parent 9 (means = 100.85 and 1008.85) is $D^2_{79} = 0.1$.

The method can also show which traits contribute most to the genotypic diversity among the potential parents. Bhatt[3] observed that kernel weight contributed most of the genetic diversity among the 40 wheat genotypes. In the absence of correlations between traits within genotypes, the generalized distance procedure will place the greatest emphasis on the trait with the lowest within-genotype variability (i.e., environmental variation). This will generally be the trait with the highest heritability.

The emphasis in Bhatt's[3,7] application of generalized distance was to identify parents that were genotypically diverse. This contrasts with the application of vector and canonical analysis where the purpose was to identify parents whose mid-parent values were as close as possible to a specified ideal. Only when two or more crosses are expected to produce populations with similar means, would one be concerned about genotypic variability, unless one is interested in initiating a long-term recurrent selection program.

As with canonical analysis, the generalized distance measurement may put undue emphasis on traits which are not really important in the breeding program.

IV. SUMMARY

The multivariate statistical methods used for choosing parents for a plant breeding program serve primarily to reduce the number of variables to one or two. Beyond this obvious simplification, the methods group naturally into those in which parents are chosen because their mid-parent value is as close as possible to the specified objective of the breeding program, and those where parents are chosen as to maximize the genotypic variation within the progeny population. The latter approach may be most suitable for long-term recurrent selection, while the former is to be preferred for short-term breeding programs.

The methods also can be grouped according to the ability of the plant breeder to specify relative importance for the various traits. Vector analysis and least-squares analysis provide an opportunity for specifying that some traits are more important than others in a breeding program. In contrast, the importance of various traits in canonical analysis, and in the generalized distance method, is entirely dependent upon the relative heritabilities of the traits and the relationships among them. Caution is required in choosing traits to be included in the analysis if one anticipates using either of these methods.

Application of any of these methods to a breeding program will require some understanding of the multivariate statistical techniques used. The examples discussed in this chapter should provide some understanding of the methods, as well as serve to emphasize some of the difficulties that can be encountered.

REFERENCES

1. **Grafius, J. E.,** A Geometry of Plant Breeding, Michigan State University Agricultural Experiment Station Res. Bull. No. 7, East Lansing, 1965.
2. **Whitehouse, R. N. H.,** Canonical analysis as an aid in plant breeding, in Barley Genetics II, Proc. 2nd Int. Barley Genetics Symp., 1970, 269.
3. **Bhatt, G. M.,** Comparison of various methods of selecting parents for hybridization in common bread wheat *(Triticum aestivum* L.), *Aust. J. Agric. Res.,* 24, 457, 1973.
4. **Pederson, D. G.,** A least-squares method for choosing the best relative proportions when intercrossing cultivars, *Euphytica,* 30, 153, 1981.
5. **Whitehouse, R. N. H.,** An application of canonical analysis to plant breeding, in Proc. 5th Cong. Eucarpia, Milano, 1968, 61.
6. **Riggs, T. J.,** The use of canonical analysis for selection within a population of spring barley, *Ann. Appl. Biol.,* 74, 249, 1973.
7. **Bhatt, G. M.,** Multivariate analysis approach to selection of parents for hybridization aiming at yield improvement in self-pollinated crops, *Aust. J. Agric. Res.,* 21, 1, 1970.

Chapter 7

RECOMMENDATIONS FOR USING SELECTION INDICES IN PLANT BREEDING

I. INTRODUCTION

Preceding chapters have been based on a review of literature, and have included discussions of the advantages of using selection indices for the simultaneous improvement of two or more traits, as well as for enhancing response to selection for a single trait. The limitations of selection indices, particularly in relation to the estimation of population parameters and the use of poor estimates, were discussed. The purpose of this chapter is to assess current knowledge and develop recommendations about when one should use a selection index and what type should be used.

II. GENERAL CONSIDERATIONS

A. Are Selection Indices Necessary in Plant Breeding?

This question is motivated by the sparsity of documentation showing that selection indices are actually being used in crop improvement. Although the methods have been known for decades, they seem to be more popular in animal improvement programs than in plant improvement programs. For single trait selection, selection indices are required only if one wishes to combine data from different stages of testing or from different relatives in the same test. This application does not seem to be particularly well developed in relation to plant populations. Recent papers, such as those by Cunningham[1] and by Moreno-Gonzalez and Hallauer,[2] may signify a growing interest in this type of application.

Multiple trait selection can be facilitated through index selection, independent culling, or tandem selection. Evaluations of these three methods suggest that index selection is most effective in all cases, and decidedly more effective than independent culling or tandem selection in some cases. However, these evaluations all require that relative economic weights be known. If one were to use some other criterion than improvement in overall genotypic worth, index selection might not always be most efficient. With linear worth functions, selection indices are expected to be more effective than tandem selection or independent culling for improving two or more traits in a recurrent selection program. Their limitations for this purpose will be discussed in a later section of this chapter.

In short-term selection programs (one or few cycles), the assessment of relative importance of different traits is often compromised in recognition of the limited material in the breeding program. Formalized selection indices will not likely be useful for multiple trait selection in these types of breeding programs. However, index methodology could be used to great advantage in choosing parents for these breeding programs.

Selection indices are required to objectively combine data from various stages of testing or from different relatives within a population; they may be used to enhance improvement in multiple traits subjected to recurrent selection; and they show considerable promise for choosing parents for breeding programs. Because index selection can be applied to these critical aspects of crop improvement, plant breeders should be familiar with their advantages and disadvantages, their limitations, and the basic methodology of their use.

B. How Should One Choose Traits to be Included in a Selection Index?

One of the requirements for efficient index selection is that estimates of population variances and covariances be sufficiently precise to give the expected response to selection.

Table 1
VALUES OF THE PARTIAL CORRELATION $r_{YB.A}$
FOR VARIOUS LEVELS OF r_{AB}

				r_{AB}		
r_{YA}	r_{YB}	0.0	0.2	0.4	0.6	0.8
0.3	0.3	0.31	0.26	0.21	0.16	0.10
0.3	0.7	0.73	0.68	0.66	0.68	0.80
0.7	0.3	0.98	0.80	0.64	0.49	0.33
0.7	0.7	0.42	0.22	0.03	−0.21	−0.61

While it is difficult to quantify the degree of precision required for developing useful indices, it is known that estimates will be less precise if sample sizes (particularly the number of genotypes tested) are small, if the number of traits included in the index is large, and/or if the traits in the index include one or more groups of highly correlated traits (see Chapter 5). One of the first steps in developing a selection index should be a thorough evaluation of traits to be included in the index.

Evaluation and choice of candidate traits for a selection index might proceed in the following way. If two or more traits are highly correlated, it is doubtful that more than one should be included in the index. In the first place, it is known that the sampling variance of estimated covariances will be higher for pairs of traits with high correlations. For example, the variance of the estimated covariance[3] between traits 1 and 2 is

$$\frac{1}{n} [\sigma_{12}\sigma_{12} + \sigma_1^2\sigma_2^2]$$

where σ_{12}, σ_1^2 and σ_2^2 are the covariance between traits 1 and 2, and variances of the two traits, which are each estimated with n degrees of freedom. Upon rearrangement, the variance of the estimated covariance is

$$\frac{1}{n} [(1 + r^2)\,\sigma_1^2\sigma_2^2]$$

which will increase as the correlation, r, approaches −1 or +1. For this reason, one should expect larger sampling errors when working with groups of highly correlated traits.

A second reason for not including all of a group of highly correlated traits is that the addition of such traits to an index may contribute little to overall response to selection. If the partial correlation between a particular candidate trait and genotypic worth is small, discarding that trait from the index can sometimes lead to improvement in response by lessening the impact of sampling errors.[4] Partial correlations are expected to be smaller if two or more traits are highly correlated. For example, if Y indicates the genotypic value of a trait to be improved by index selection, and if A and B refer to the phenotypic values of two traits being considered for inclusion in the index, the partial correlation between Y and B, given that trait A is constant, is

$$r_{YB.A} = \frac{r_{YB} - r_{YA}r_{AB}}{[(1 - r_{YA}^2)(1 - r_{AB}^2)]^{0.5}}$$

The effect of the correlation between traits A and B on the partial correlation between Y and B is shown in Table 1. If both traits are approximately equally correlated with Y (r_{YA}

= r_{YB} = 0.3 or 0.7), increases in the correlation between A and B will result in decreases in the partial correlation between either of the traits and Y. If trait B has a lower simple correlation with Y than trait A, its partial correlation rapidly approaches zero or a negative value as the correlation between A and B increases. If trait B is more highly correlated with Y than is trait A, its partial correlation with Y will decrease to a minimum, and then increase as r_{AB} increases from 0.0. In this case, the covariance between Y (adjusted for linear regression on trait A) and B (adjusted for linear regression on trait A) does not decrease as fast as the standard deviation of B (adjusted for linear regression on trait A).

While this discussion suggests that traits having low partial correlations with genotypic worth should be discarded from the index, it is not possible to give specific recommendations as to whether or not a particular trait should be discarded. Cochran[4] gave arguments which support the practice of discarding traits with low partial correlation coefficients. However, it was pointed out that this practice is itself subject to error. Moreover, in practical applications, it is difficult to decide if a trait has a partial correlation large enough to warrant its inclusion in an index.

The important point is to recognize that inclusion of highly correlated variables in an index may lead to very little improvement in selection response because of small partial correlations, and may even lead to reduced response because of increased sampling error. Traits should be included in a selection index only if they can be expected to have a nontrivial correlation with genotypic worth and are expected to provide information about genotypic worth that is not already supplied by another trait.

Particular caution should be used when dealing with traits that are mathematically derived from measurements on primary traits. An example involves the use of leaf area, leaf weight, and specific leaf weight in an index for improving seed yield of *Brassica*.[5] In this instance, leaf area and leaf weight can be expected to be highly correlated. Moreover, specific leaf weight is the ratio of leaf weight to leaf area. It is unlikely that any advantage can be obtained by including more than two of these three variables in any selection index. In fact, since leaf area and leaf weight were highly correlated (r = 0.80), the best index may include only leaf area or leaf weight. Based on phenotypic correlations with seed yield, the partial correlation between leaf weight and seed yield, both adjusted for linear regression on leaf area, is

$$[0.601 - (0.586)(0.800)]/[(1 - 0.586^2)(1 - 0.800^2)]^{0.5} = 0.272$$

Similarly, the partial correlation between seed yield and leaf area, both adjusted for leaf weight, is 0.219. On this basis, one might choose to include leaf weight in the index and to question whether or not inclusion of leaf area would be worthwhile.

Estimated response to selection will increase with each new trait that is included in an index. However, estimates of expected response are biased upwards and fail to reflect the consequences of errors in the index coefficients. In choosing traits to be included in an index, a certain amount of subjectivity is required to temper decisions based on sample estimates with knowledge of the sources, nature, and consequences of errors in those estimates.

C. Should a Weight-Free Index Be Used Instead of an Estimated Optimum Index?

In the absence of estimates of genotypic and phenotypic variances and covariances, the need for an objective rule for selection can sometimes be met by using the multiplicative index proposed by Elston.[6] Baker[7] has suggested that this index can be approximated by using a base index in which the weights of individual traits are set equal to the reciprocals of their respective phenotypic standard deviations. The method may be quite useful when estimates of population parameters are not available, and when the relative importance of traits is judged to be equal in the sense that a change of one phenotypic standard deviation is of equal importance for all traits. When traits are not of equal value in this sense, it may be better to use the base index developed by Williams.[8]

While a lack of estimates of genotypic and phenotypic variances and covariances necessitates the use of a weight-free index or a base index, a far more fundamental question is whether or not selection responses from estimated indices are sufficiently greater than those from a base index to warrant the effort required to obtain estimates of population parameters. Williams,[8] in discussing the choice of selection indices, suggested that an estimated index should be used only if it provides worthwhile improvement over any alternate index. However, if the base index is easier to construct and if it gives results as good as or better than the estimated index, the base index is to be preferred.

Williams[9] suggested that the value of an optimum index (based on known population parameters), relative to a base index, could be measured by the ratio of their respective correlations with genotypic worth. If

$$W = a_1G_1 + a_2G_2 + \ldots + a_nG_n$$

is the genotypic worth and if optimum and base indices are given by

$$I_o = b_1P_1 + b_2P_2 + \ldots + b_nP_n \text{ and}$$

$$I_b = a_1P_1 + a_2P_2 + \ldots + a_nP_n$$

respectively, the relative efficiency (RE) of the two indices is

$$RE = \frac{r_{WIo}}{r_{WIb}}$$

where r_{WIo} is the correlation between genotypic worth and the optimum index, and r_{WIb} is the correlation between genotypic worth and the base index. In matrix notation,

$$RE = \frac{[a'GP^{-1}Ga]^{0.5} \, [a'Pa]^{0.5}}{[a'Ga]} \tag{3}$$

where $a = [a_1 \; a_2 \; \ldots \; a_n]$ is the vector of relative economic weights of the n traits, G is an $n \times n$ matrix of population genotypic variances and covariances, and P is an $n \times n$ matrix of population phenotypic variances and covariances. This expression is quite complex when written in long-hand notation. For n traits, the evaluation of the expected relative efficiency of an optimum index and a base index requires knowledge of n relative economic values, n genotypic variances, n heritabilities, $n(n - 1)/2$ genotypic correlations, and $n(n - 1)/2$ phenotypic or environmental correlations. Either phenotypic or environmental correlations may be used by applying the relationship given by Falconer,[10] which shows that

$$r_{P(ij)} = r_{G(ij)}h_ih_j + r_{E(ij)}[(1 - h_i^2)(1 - h_j^2)]^{0.5}$$

Genotypic covariances can be expressed as

$$\sigma_{G(ij)} = r_{G(ij)}\sigma_{G(i)}\sigma_{G(j)}$$

phenotypic variances as

$$\sigma_{P(i)}^2 = \frac{\sigma_{G(i)}^2}{h_i^2}$$

and phenotypic covariances as

$$\sigma_{P(ij)} = r_{P(ij)}\sigma_{P(i)}\sigma_{P(j)} = r_{P(ij)}\frac{\sigma_{G(i)}}{h_i}\frac{\sigma_{G(j)}}{h_j}$$

Equation 3 can be easily evaluated under certain simplifying assumptions. If the n traits are independent, $(r_{P(ij)} = r_{G(ij)} = 0)$, equation 3 becomes

$$RE = \frac{[\sum a_i^2\sigma_{G(i)}^2 h_i^2]^{0.5} [\sum a_i^2\sigma_{G(i)}^2/h_i^2]^{0.5}}{\sum a_i^2\sigma_{G(i)}^2}$$

Furthermore, if h_i^2 is constant for all traits,

$$RE = \frac{[h^2\sum a_i^2\sigma_{G(i)}^2]^{0.5} \left[\frac{1}{h^2}\sum a_i^2\sigma_{G(i)}^2\right]^{0.5}}{\sum a_i^2\sigma_{G(i)}^2}$$

$$= [h^2]^{0.5}\left[\frac{1}{h^2}\right]^{0.5} = 1.0$$

Thus, for independent traits, the optimum index will be no more efficient than a base index if all traits have the same heritabilities.

If all traits are independent and equally important in the sense that $a_i\sigma_{G(i)} = k$ for all i, then

$$RE = \frac{[k\sum h_i^2]^{0.5} [k\sum 1/h_i^2]^{0.5}}{kn} = \frac{1}{n}[\sum h_i^2\sum 1/h_i^2]^{0.5}$$

In this case, the greatest advantage of the optimum index over a base index will occur when half the traits have maximum heritability, and the remaining half have minimum heritability. Then,

$$RE = \frac{1}{n}\left[\frac{n}{2}h_{max}^2 + \frac{n}{2}h_{min}^2\right]\left[\frac{n}{2}\frac{1}{h_{max}^2} + \frac{n}{2}\frac{1}{h_{min}^2}\right]$$

$$= \frac{1}{4}[h_{max}^2 + h_{min}^2][1/h_{max}^2 + 1/h_{min}^2]$$

This ratio is now independent of the number of traits included in the index. Values of this ratio (Table 2) show that the maximum advantage of an optimum index over a base index for independent, equally important traits will occur when at least one trait has very low heritability. If the minimum heritability is equal to or greater than 0.3, the maximum advantage of an optimum index is expected to be about 15%.

An evaluation of more general conditions was carried out by evaluating Equation 3 for various combinations of parameters specific to the case of two traits (Table 3). For two traits with equal importance $(a_1\sigma_{G(1)} = a_2\sigma_{G(2)})$, the greatest advantage of the optimum index over a base index occurs when heritabilities differ markedly. These advantages are enhanced as the dependence between the traits increases. The advantage of the optimum index is greatest when both traits have equal importance, though large discrepancies between genotypic and environmental correlations do cause some minor exceptions to this trend.

Table 2
RELATIVE EFFICIENCY OF AN
OPTIMUM INDEX RELATIVE TO
A BASE INDEX WITH
INDEPENDENT TRAITS[a] OF
EQUAL WEIGHT[b]

h^2_{max}	h^2_{min}	Relative efficiency (%)
0.9	0.9	100
0.9	0.7	101
0.9	0.5	104
0.9	0.3	115
0.9	0.1	167
0.7	0.7	100
0.7	0.5	101
0.7	0.3	109
0.7	0.1	151
0.5	0.5	100
0.5	0.3	103
0.5	0.1	134
0.3	0.3	100
0.3	0.1	115
0.1	0.1	100

[a] $r_P = r_G = O$
[b] $a_i \sigma_{G(i)} = k$

Based on the evaluation of Equation 3 for two traits and, with simplifying assumptions, for an arbitrary number of traits, an optimum index would be expected to have important advantages over a base index only if one or more, but not all, of the traits in the index were characterized by heritabilities less than about 30% or when traits are highly correlated.

Williams[8,9] emphasized that an estimated index can never be as efficient as an optimum index. In fact, if estimates of population parameters are not reliable, an estimated index may actually give less response on average than a base index. Since response to an estimated index will be less than indicated by Equation 3, there is a strong argument against using an estimated index except in those cases where the advantage of an optimum index over a base index is substantial, and when good estimates are available for the required population parameters.

Another way of evaluating the advantage of an optimum index over a base index is to use Henderson's[11] approach to the development of the optimum selection index. Henderson showed that the development of an index for the improvement of several traits is equivalent to developing a separate index for each trait and then weighting each index according to the relative economic importance of the corresponding trait. In this context, Williams' base index[8] is equivalent to the case where individual indices are composed only of the trait to be improved by that index. Thus, if the genotypic value of trait i, G_i, is to be improved by using the index $I_i = h^2_i P_i$, and if all h^2_i are equal, the method given by Henderson[11] is equivalent to the base index given by Williams.[8] In this case, the base index is

$$I_b = a_1 P_1 + a_2 P_2 + \ldots + a_n P_n$$

while the Henderson optimum index is

$$I_h = a_1 h^2_1 P_1 + a_2 h^2_2 P_2 + \ldots + a_n h^2_n P_n$$

Table 3
RELATIVE EFFICIENCY (%) OF AN OPTIMUM SELECTION INDEX FOR TWO TRAITS OVER A BASE INDEX

h_1^2	h_2^2	$r_G = -0.5$ $r_E = -0.5$	$r_G = -0.5$ $r_E = 0.0$	$r_G = -0.5$ $r_E = 0.5$	$r_G = 0.0$ $r_E = 0.0$	$r_G = 0.0$ $r_E = 0.5$	$r_G = 0.5$ $r_E = 0.5$
		$(a_1\sigma_{G(1)}/a_2\sigma_{G(2)} = 1)$					
0.9	0.9	100	100	100	100	100	100
0.9	0.5	105	105	105	104	104	105
0.9	0.1	172	168	167	167	168	172
0.5	0.5	100	100	100	100	100	100
0.5	0.1	141	135	136	134	138	141
0.1	0.1	100	100	100	100	100	100
		$(a_1\sigma_{G(1)}/a_2\sigma_{G(2)} = 2)$					
0.9	0.9	100	100	101	100	100	100
0.9	0.5	103	106	110	103	105	104
0.9	0.1	161	169	180	146	154	146
0.5	0.5	100	103	112	100	101	100
0.5	0.1	125	142	170	123	140	129
0.1	0.1	100	110	140	100	104	100
		$(a_1\sigma_{G(1)}/a_2\sigma_{G(2)} = 4)$					
0.9	0.9	100	100	101	100	100	100
0.9	0.5	100	102	106	101	103	102
0.9	0.1	116	125	135	118	126	122
0.5	0.5	100	104	115	100	103	100
0.5	0.1	103	117	143	108	126	117
0.1	0.1	100	113	151	100	109	100

Whether or not the base index should be used depends, therefore, on whether or not heritability is constant for all traits and whether or not selection response for any of the traits with nonzero economic value can be improved by including secondary traits in the corresponding selection index.

A guideline as to when to use an estimated index can therefore be based on a consideration of when a selection index will give greater response for a single trait than will direct selection on that trait alone. Consider selection for improved genotypic values of trait 1 (i.e., G_1), based on its own phenotypic value (P_1), or on the basis of an index $I = b_1P_1 + \ldots + b_nP_n$. The response to index selection relative to direct selection is expected to equal

$$\frac{r_{G(1)I}}{r_{G(1)P(1)}} = \frac{R_{G(1).P(1)P(2)\ldots P(n)}}{r_{G(1)P(1)}}$$

where $R_{G(1).P(1)P(2)} \cdots$ is the multiple correlation between G_1 and the phenotypic values of the traits included in the selection index. If this ratio is sufficiently greater than unity for one or more traits with economic importance, an estimated index should be used in place of a base index. For an index based on two traits, $I = b_1P_1 + b_2P_2$, the ratio of $r_{G(1)I}$ to $r_{G(1)P(1)}$ for optimum values of b_1 and b_2 is

$$\left[\frac{R^2_{G(1).P(1)P(2)}}{r^2_{G(1)P(1)}} \right]^{0.5} =$$

$$\left[\frac{r^2_{G(1)P(1)} + r^2_{G(1)P(2).P(1)} - r^2_{G(1)P(2).P(1)} \, r^2_{G(1)P(1)}}{r^2_{G(1)P(1)}} \right]^{0.5} =$$

$$\left[1 + \frac{(1 - r^2_{G(1)P(1)}) \, (r^2_{G(1)P(2).P(1)})}{r^2_{G(1)P(1)}} \right]^{0.5}$$

where $r_{G(1)P(1)}$ is the correlation between genotypic and phenotypic values for trait 1 ($= h_1$), and $r_{G(1)P(2) \, . \, P(1)}$ is the partial correlation between G_1 and P_2 after both have been adjusted for linear regression on P_1.

Increased efficiency of an optimum index over a base index will require that the ratio $(1 - r^2_{G(1)P(1)}/r^2_{G(1)P(1)})$ be high, and that the partial correlation $r_{G(1)P(2) \, . \, P(1)}$ also be high. The first term is equal to $(1 - h_1^2/h_1^2)$ and will be high if h_1^2; the heritability of the economically important trait, is low. Thus, the necessary conditions for choosing an optimum index over a base index are that the heritability of at least one trait with economic importance be low, and that there exists at least one secondary trait which has a high partial correlation with the genotypic value of the primary trait.

Many economically important traits have sufficiently low heritabilities to meet the above requirement. The critical question again lies in the size of the partial correlation between phenotypic value of a secondary trait and the genotypic value of the primary trait. A high partial correlation requires that the phenotypic values of the two traits not be highly correlated, but that the phenotypic value of the secondary trait be highly correlated with the genotypic value of the primary trait. These conditions will be met if there is a strong genotypic correlation between the two traits, and if the primary trait has low heritability while the secondary trait has high heritability. Using the data of Thurling,[5] the heritability of leaf weight (0.63) exceeded that for seed yield (0.55), and there was a high genotypic correlation between the two traits (0.85). However, the estimated advantage of using an index based on yield and leaf weight was only 7% over direct selection for yield alone. Unless the true genotypic correlation and/or the difference in heritabilities is much greater than was estimated, it is doubtful that the advantage is sufficiently great to warrant using an index. With such a small advantage, the plant breeder runs the risk that the estimated index will actually be less efficient than a base index, or even less efficient than single trait selection for seed yield.

This discussion suggests that, where heritabilities differ markedly, one might consider using a modified base index in which weights are taken to be the product of the relative economic values and the heritabilities of the individual traits. If relative heritabilities were known with good precision, application of Equation 3 shows that such an index would be expected to give responses intermediate to those expected from base and optimum indices, and generally much closer to that expected of an optimum index. However, the impact of errors in the estimation of heritability has not been evaluated.

Smith et al.[12] proposed an index in which the estimated heritabilities were used as index coefficients. In their example, relative economic weights were the same, except for sign, for all traits. A more efficient index, when both heritabilities and relative economic weights are known to vary from trait to trait, would be

$$I = a_1 h_1^2 P_1 + a_2 h_2^2 P_2 + \ldots + a_n h_n^2 P_n$$

where each trait is weighted according to the product of its heritability and its relative economic value.

D. Should a Restricted Selection Index Be Used?

Restricted selection indices are designed to result in specified responses in one or more traits after completing a number of cycles of index selection. There are two instances where these types of indices should be considered. When it is difficult to develop sound definitions of relative economic values of individual traits, it may be practical to specify selection goals in terms of optimum values for each trait. Restricted selection indices could be used to shift the population toward those desired values.

A second situation where it may be necessary to use restricted selection indices is where certain traits must be maintained at current population levels during the selection process. One would try to develop an index which would maintain those traits at their starting levels.

In addition to the usual problems concerning choice of traits to be included in an index, restricted indices have the added problem that restrictions will reduce overall response to selection. If restrictions are poorly chosen, the attempt to obtain specified responses in those traits may severely limit overall response to selection. This can be demonstrated by considering two traits which have equal economic values, but which are negatively correlated and have different heritabilities. For example, if heritabilities are $h_A^2 = 0.6$ and $h_B^2 = 0.3$ for traits A and B, respectively, if genotypic and phenotypic correlations between the two traits are -0.6 and -0.2, and if the phenotypic variances are 1.0 for both traits, the optimum index would be $I_o = P_A + 0.21\ P_B$. If, on the other hand, one were to develop a restricted index designed to give equal amounts of response in both traits, the resulting index would be $I_r = P_A + 1.64\ P_B$. Response in total genotypic worth $(G_A + G_B)$ would be expected to be 17% less with the restricted index than with the unrestricted index. The requirement for equal responses in both traits would sacrifice response in more heritable trait A in order to obtain equal response in trait B.

If one has decided to use a restricted index, the choice of a particular method becomes a confusing task. Of necessity, the presentation of the various restricted index methods is usually done in terms of matrix algebra. Often, matrix algebra is not well understood by those who would apply these methods to crop improvement. Harville[13] indicated that the method he proposed for developing restricted indices should be more efficient than the one developed earlier by Tallis.[14] Tallis (personal communication, 1984) indicated that the Harville[13] index should be no more efficient than the Tallis[14] index. The Harville[13] index can be applied to cases where the index includes traits which are not restricted, or where the index includes only traits for which proportionality constraints are specified. In the latter case, the Harville[13] method is identical to a specific situation covered by the Tallis[14] index, as well as to the method of "desired gains" developed by Pesek and Baker.[15] The former case, where unconstrained as well as constrained traits are included in the index, gives identical results whether one uses the Harville[13] approach or the method developed independently by Tai[16] and also discussed by Yamada et al.[17] Recognizing the relationships among the various methods leads to the conclusion that the method described by Harville[13] should be the method of choice for most cases of restricted indices.

III. USE OF SELECTION INDICES

A. Nonrecurrent Selection Programs

Much of the effort in breeding self-pollinated species involves a single cycle of selection within a particular cross. Selection indices will not likely prove useful in this type of breeding program, except to combine data from various stages of testing, or to combine data from different relatives in single trait selection. For multiple traits, perceived economic values will often depend more upon the observed values of the genotypes in the breeding program than on actual economic importance in the crop. Since the breeding program is a closed program, the breeder will try to choose the best material from that available. Since that choice may not accurately reflect the economic importance of various traits, it is unlikely that formal index selection would be useful.

For single trait selection in two or more stages, the principles of index selection may be helpful. Cochran[4] considered two stages of selection for a single trait. A fraction p of the genotypes would be selected in the first stage, and each selected genotype would be replicated 1/p as many times in the second stage as in the first stage. Final selection would be based on weighted data from both stages. For this particular type of program, Cochran suggested that one should select approximately 20% of the population in the first stage. Cunningham[1] also considered multistage testing.

Similar, but somewhat more complex, approaches might be used when data on each genotype include data from one location in one season and from several locations in another. Since only one trait is being considered, relative economic values are not a consideration. The optimum weighting of the data from the various stages will depend upon the relative sizes of genotypic, environmental, and genotype-environmental variances. The general methods used in developing selection indices can be used in developing appropriate weights for the data collected in different stages.

In some breeding programs, there will be a family structure to the genotypes being tested at any particular stage. For example, F_5-derived lines might be grouped into F_3 subpopulations. In cases such as this, index methods, such as those discussed by Falconer,[10] could be used for weighting family means and individual observations. Often, a near-optimum index can be obtained through a theoretical assessment of the family structure and of anticipated relationships based on simplified genetic assumptions.

B. Recurrent Selection

In recurrent selection programs where more than one trait contributes to the overall worth of a genotype, an objective index is essential for consistent selection. Whether the index should be a weight-free index such as that proposed by Elston,[6] a base index as recommended by Williams,[8,9] a restricted index,[13-16] or an estimated optimum index, will depend largely upon the general considerations presented in Section II. If the relative economic values cannot be specified, a weight-free index or a restricted index may be a good first choice. If relative economic importance can be specified, and if there is no need to restrict the change in any particular trait, a base index could enhance response in overall genotypic worth. When traits vary in heritability and show high levels of interdependence, it will be advantageous to develop good estimates of population parameters and to use an estimated optimum index.

Selection for single traits in a recurrent selection program can also require the use of a selection index to combine data from various traits or to combine data from a single trait measured on various relatives. Moreno-Gonzalez and Hallauer[2] proposed a selection index for combining full-sib and S_2 family data in full-sib reciprocal recurrent selection.

IV. PARENTAL EVALUATION IN SELF-POLLINATED SPECIES

The use of multivariate techniques for choosing parents was reviewed in Chapter 6. It was shown that there are two basic strategies in choosing parental material. It was suggested that methods for choosing parents whose mid-parent values agree most closely with specified breeding objectives are most useful for short-term breeding programs. Methods differ in how they assess the closeness of mid-parent values to the specified objectives. Rather than having this assessment based solely on the statistical method, the variance-covariance structure of the parental population, and sampling error, it would seem desirable to include some assessment of the relative importance of each trait. The methods of Grafius[18] and Pederson[19] do allow a plant breeder to include an assessment of relative importance of each trait. Either of these methods would be desirable for establishing the relative importance to be attached to the various traits. Since Grafius' method[18] for choosing potential parents for a breeding program is based on the concepts of geometry, and since these concepts are often not well understood by plant breeders, the methods described by Pederson[19] may be more useful in practice.

The requirement for being able to include some measure of relative economic importance would also seem to be important when selecting parents for genotypic diversity. It does not make sense to choose parents which are genotypically diverse for unimportant traits as a basis for a long-term recurrent selection program. Unfortunately, the methods developed thus far for identifying genotypically diverse parents have failed to incorporate any measure of relative economic importance of different traits.

Because final response to selection is determined by the choice of parental material, it is critical to choose those parents which have the proper balance of all economically important traits. Index methodology can play an important role in this step in the breeding program, particularly in short-term breeding programs.

V. DETERMINING RELATIVE ECONOMIC WEIGHTS

In most methods of index selection, one must specify the relative importance of the different traits to be included in the index. This is often a difficult task. Pesek and Baker[15] sought to circumvent this requirement by developing an estimated index which would give responses in individual traits proportional to those specified as desirable by the plant breeder. This approach is the basis of the restricted indices described by Tallis[14] and Harville.[13]

There is reason to question whether or not the use of restricted indices really circumvents the need for specifying relative economic importance. Perhaps the method merely disguises the problem in another form. For independent traits, the method of Pesek and Baker[15] will give index coefficients $b_i = d_i/\sigma_{G(i)}$, where $\sigma_{G(i)}$ is the genotypic standard deviation for trait i and d_i is the desired response for that trait. An optimum index for independent traits would require $b_i = a_i h_i^2$. Thus, if $a_i h_i^2 = d_i \sigma_{G(i)}^2$ or $d_i = a_i \sigma_{G(i)}^2 h_i^2$, the restricted index would give the same results as the optimum index. In effect, at least for independent traits, specifying desired gains (or optimum responses) is equivalent to specifying relative economic values except for a factor which is dependent upon heritability and genotypic variance.

It is reasonably easy to specify relative economic values for traits which relate to different crop products. Eagles and Frey[20] assigned relative economic weights to oat, *Avena sativa*, grain yield and straw yield based on the relative value of these two products in the marketplace. In oil crops, it should be possible to assign relative economic weights to oil and protein concentrations by evaluating the market value of oil and of protein meal. Assigning relative economic weights is much more difficult when two or more traits relate to different aspects of one product. Protein concentration, flour yield, and grain yield are all important attributes of bread wheat. The first two are related to the quality of the product while yield is related to the amount of the product. It is quite easy to justify the need for increased yield and to put an economic value on such an increase. On the other hand, it is extremely difficult to attach economic values to increases in protein concentration or flour yield. Determining economic value is particularly difficult if some characteristics must meet certain minimum standards before a new genotype can even be considered for commercial production.

One way of developing ad hoc estimates of relative economic importance is that proposed by Grafius[18] and discussed in Chapter 6. By having one or more knowledgable persons give an overall worth score to each of a number of genotypes, one can then use multiple regression to determine the relative importance of each trait in contributing to the overall worth score. This approach might be a good first step in trying to determine relative importance of various traits in a breeding program. Pederson[19] developed the concept of specifying optimum values and preferred maximum deviations for each trait. Pederson discussed the use of these two estimates in choosing parents for a breeding program. While this approach could be used in developing estimates of relative importance of various traits, it is not clear how such estimates could be incorporated into an optimum selection index.

Estimates of relative economic importance must be related to the units of measurement of each trait. If the units of measurement are changed, the relative economic value must be changed. Thus, if an increase of 1 t/ha in grain yield is twice as important as an increase of 1% in protein concentration, it is important to realize that the relative economic values are 2/t/ha and 1/% or, equivalently, 0.002/kg/ha and 1/%.

VI. SUMMARY

Questions which a plant breeder would ask about selection indices before applying them to a breeding program have been discussed in this chapter. For single traits, index methods can be used to combine data from different stages of testing or from relatives evaluated for the same trait. These methods will be useful in single cycle selection as well as for recurrent selection programs. For multiple trait selection, selection indices should be useful as an aid for choosing parents particularly in self-pollinated species. Selection indices should also be more effective than independent culling and tandem selection for improving multiple traits in recurrent selection programs.

Care must be taken in choosing secondary traits for inclusion in a selection index. There is little advantage, and may be considerable disadvantage, to including more than one of a group of highly correlated traits. If heritabilities of all traits included in an index are intermediate to high, a base index, in which each trait is weighted according to its relative economic value, may be the best index to use. If traits differ markedly in heritability, a modified base index in which weights are set equal to the product of relative economic value and estimated heritability should be better. In any application of multiple trait selection, the development of realistic estimates of relative economic importance presents an important but difficult task.

REFERENCES

1. **Cunningham, E. P.,** Multi-stage index selection, *Theor. Appl. Genet.,* 46, 55, 1975.
2. **Moreno-Gonzalez, J. and Hallauer, A. R.,** Combined S_2 and crossbred family selection in full-sib reciprocal recurrent selection, *Theor. Appl. Genet.,* 61, 353, 1982.
3. **Tallis, G. M.,** Sampling errors of genetic correlation coefficients calculated from analyses of variance and covariance, *Aust. J. Stat.,* 1, 35, 1965.
4. **Cochran, W. G.,** Improvement by means of selection, in Proc. 2nd Berkeley Symp. Math. Stat. Prob., 1951, 449.
5. **Thurling, N.,** An evaluation of an index method of selection for high yield in turnip rape, *Brassica campestris* L. ssp. *oleifera* Metzg., *Euphytica,* 23, 321, 1974.
6. **Elston, R. C.,** A weight-free index for the purpose of ranking or selection with respect to several traits at a time, *Biometrics,* 19, 85, 1963.
7. **Baker, R. J.,** Selection indexes without economic weights for animal breeding, *Can. J. Anim. Sci.,* 54, 1, 1974.
8. **Williams, J. S.,** The evaluation of a selection index, *Biometrics,* 18, 375, 1962.
9. **Williams, J. S.,** Some statistical properties of a genetic selection index, *Biometrika,* 49, 325, 1962.
10. **Falconer, D. S.,** *Introduction to Quantitative Genetics,* 2nd ed., Longman, New York, 1981.
11. **Henderson, C. R.,** Selection index and expected genetic advance, in Statistical Genetics and Plant Breeding, No. 982, Hanson, W. D. and Robinson, H. F., Eds., National Academy of Sciences, National Research Council, Washington, D.C., 1963, 141.

12. **Smith, O. S., Hallauer, A. R., and Russell, W. A.,** Use of index selection in recurrent selection programs in maize, *Euphytica,* 30, 611, 1981.
13. **Harville, D. A.,** Index selection with proportionality constraints, *Biometrics,* 31, 223, 1975.
14. **Tallis, G. M.,** A selection index for optimum genotype, *Biometrics,* 18, 120, 1962.
15. **Pesek, J. and Baker, R. J.,** Desired improvement in relation to selection indices, *Can. J. Plant Sci.,* 49, 803, 1969.
16. **Tai, G. C. C.,** Index selection with desired gains, *Crop Sci.,* 17, 182, 1977.
17. **Yamada, Y., Yokouchi, K., and Nishida, A.,** Selection index when genetic gains of individual traits are of primary concern, *Jpn. J. Genet.,* 50, 33, 1975.
18. **Grafius, J. E.,** A Geometry of Plant Breeding, Michigan State University Agricultural Experiment Station Res. Bull. No. 7, 1965.
19. **Pederson, D. G.,** A least-squares method for choosing the best relative proportions when intercrossing cultivars, *Euphytica,* 30, 153, 1981.
20. **Eagles, H. A. and Frey, K. J.,** Expected and actual gains in economic value of oat lines from five selection methods, *Crop Sci.,* 14, 861, 1974.

Chapter 8

DEVELOPMENT AND USE OF SELECTION INDICES

I. INTRODUCTION

In this chapter, the methods required to develop and apply selection indices to plant breeding programs are reviewed and the required calculations are outlined by using simulated data. While it is not possible to anticipate all possible applications, the material covered herein should provide a good base for most situations. In matrix terminology, most selection index methods are based on the solution of the equation

$$\mathbf{Pb} = \mathbf{Ga}$$

where $\mathbf{P}$ is a matrix consisting of estimates of phenotypic variances and covariances, $\mathbf{G}$ is a matrix consisting of estimates of genotypic variances and covariances, $\mathbf{a}$ is a vector of relative economic weights, and $\mathbf{b}$ is a vector of index coefficients which are developed by solving the set of simultaneous equations. The index coefficients are estimated as

$$\mathbf{b} = \mathbf{P}^{-1}\,\mathbf{Ga}$$

where $\mathbf{P}^{-1}$ is the inverse of the phenotypic variance-covariance matrix.

When selection is designed to improve a single trait, the vector of economic weights includes zeros except for the one trait of economic importance, and $\mathbf{G}$ becomes a vector of genotypic covariances with that same trait. If this methodology is to be applied to multistage testing, $\mathbf{P}$ and $\mathbf{G}$ include estimates of phenotypic and genotypic variances and covariances between the same trait at different stages of testing. For combining data from different relatives, $\mathbf{P}$ and $\mathbf{G}$ include estimates of relationships among relatives with respect to the one trait being considered.

When reliable estimates of $\mathbf{P}$ and $\mathbf{G}$ are not available, a possible option is to use modified index coefficients such as $\mathbf{b} = \mathbf{a}$, known as a base index. In some cases, it may be more desirable to set each index coefficient equal to the product of relative economic weight and heritability of the corresponding trait.

II. SINGLE TRAIT SELECTION

A. Using Data from Correlated Traits to Enhance Response to Selection for a Single Trait

A selection index can improve single trait response to selection if one or more of the secondary traits has higher heritability than the primary trait and a high genotypic correlation with that trait (see Chapter 7). One of the examples discussed by Smith[1] concerned selection for increased grain yield of wheat, *Triticum aestivum*, plants based on an index comprised of ear number per plant, average number of grains per ear, average weight per grain, and weight of straw per plant.

A typical application of this type of index in self-pollinated crops might be the selection among partially inbred lines within a particular cross. If primary trait A is to be improved by use of an index comprised of phenotypic values of traits A, B, and C, the methods outlined in Chapter 4 (Section II) would be used to estimate the genotypic variances of each of the three traits ($\sigma^2_{G(A)}$, $\sigma^2_{G(B)}$, and $\sigma^2_{G(C)}$) the genotypic covariances among the three traits ($\sigma_{G(AB)}$, $\sigma_{G(AC)}$, and $\sigma_{G(BC)}$), the phenotypic variances of the three traits ($\sigma^2_{P(A)}$, $\sigma^2_{P(B)}$, and

$\sigma^2_{P(C)}$), and the phenotypic covariances among the three traits ($\sigma_{P(AB)}$, $\sigma_{P(AC)}$, and $\sigma_{P(BC)}$). Estimates of these parameters would then be inserted into the following simultaneous equations which would, in turn, be solved for the unknown index coefficients b_1, b_2, and b_3.

$$b_1\sigma^2_{P(A)} \; + \; b_2\sigma_{P(AB)} \; + \; b_3\sigma_{P(AC)} \; = \; \sigma^2_{G(A)}$$

$$b_1\sigma_{P(AB)} \; + \; b_2\sigma^2_{P(B)} \; + \; b_3\sigma_{P(BC)} \; = \; \sigma_{G(AB)}$$

$$b_1\sigma_{P(AC)} \; + \; b_2\sigma_{P(BC)} \; + \; b_3\sigma^2_{P(C)} \; = \; \sigma_{G(AC)}$$

Study of this set of simultaneous equations reveals that there is a pattern which, if learned, should serve to develop the necessary equations for more complex situations. Consider that the first equation is associated with trait A, the second with trait B, and the third with trait C. Subscript A occurs in every term of the first equation, subscript B occurs in every term of the second equation, and so on. Next, note that the first column to the left of the equal signs always has A as a subscript, the second column always has B as a subscript, and the third column always has C as a subscript. Furthermore, variances and covariances to the left of the equal signs are all estimates of phenotypic parameters. Where the subscript dictated by the row (equation) and by the column are the same, one uses an estimate of a phenotypic variance. If the two subscripts are different, one uses an estimate of a phenotypic covariance.

When developing a selection index to improve a single trait, there is only one column of terms to the right hand side of the equations. The subscript that is common to all those terms is the subscript of the trait that is to be improved. Only estimates of genotypic parameters occur on the right side of the equations. As with phenotypic parameters, variances are used where both subscripts are the same and covariances are used when they differ. If the trait to be improved were not included in the index, the right-hand sides of the equations would be comprised only of covariance terms.

In matrix notation, the set of simultaneous equations given above is written in short-hand notation as

$$\mathbf{Pb} \; = \; \mathbf{g}$$

where $\mathbf{P}$ is a 3 by 3 matrix of the estimated phenotypic variances and covariances, $\mathbf{b}$ is a 3 by 1 vector of unknown index coefficients, and $\mathbf{g}$ is a 3 by 1 vector of estimated genotypic variances or covariances.

If one has access to matrix handling routines on a computer, the estimated index coefficients can be solved quite easily by inserting the estimated parameters into their proper locations in matrix $\mathbf{P}$ and vector $\mathbf{g}$, and solving as $\mathbf{b} \; = \; \mathbf{P}^{-1}\mathbf{g}$. If such programs are not available, one must resort to more tedious methods for solving the particular set of simultaneous equations.

In order to demonstrate the necessary calculations, data from the genotypic and phenotypic distributions of three correlated traits were simulated by using program MNORN (Appendix 3). To use this program, one specifies the number of variables to be simulated, their genotypic variances and heritabilities, and the genotypic and environmental correlations between each pair of traits. Given this information, the program will simulate a specified sample from the multivariate normal population and print out the required genotypic and phenotypic parameters.

To demonstrate the calculations for this particular model, genotypic variances for traits A, B, and C were set equal to 100, 10, and 30, respectively, and their heritabilities were set equal to 0.4, 0.8, and 0.6, respectively. The genotypic and environmental correlations between traits A and B were 0.8 and 0.5, those between traits A and C were -0.3 and -0.4, and those between traits B and C were -0.4 and -0.4. A sample of 200 was simulated and the resulting genotypic and phenotypic covariances were as indicated by the matrices

$$\mathbf{G} = \begin{bmatrix} 87.25 & 24.16 & -16.36 \\ 24.16 & 10.23 & -7.15 \\ -16.36 & -7.15 & 30.90 \end{bmatrix}$$

and

$$\mathbf{P} = \begin{bmatrix} 213.49 & 30.49 & -32.48 \\ 30.49 & 12.12 & -9.52 \\ -32.48 & -9.52 & 51.00 \end{bmatrix}$$

In the simulated sample of 200, the genotypic variances were 87.25, 10.23, and 30.90 rather than 100, 10, and 30 as planned. Observed heritabilities for the three traits were $87.25/213.49 = 0.41$, $10.23/12.12 = 0.84$, and $30.90/51.00 = 0.61$, rather than the planned values of 0.4, 0.8, and 0.6 for traits A, B, and C, respectively. The genotypic correlations in the sample were 0.81 for A and B, -0.32 for A and C, and -0.40 for traits B and C. These compare to planned values of 0.8, -0.3, and -0.4, respectively.

Based on the estimates given above, the simultaneous equations which are to be solved in order to estimate the weightings of the three traits in the selection index are as follows.

Eq. I $\quad 213.49\, b_1 \quad + \; 30.49\, b_2 \; - \; 32.48\, b_3 \; = \quad 87.25$

Eq. II $\quad 30.49\, b_1 \quad + \; 12.12\, b_2 \; - \quad 9.52\, b_3 \; = \quad 24.16$

Eq. III $\; -32.48\, b_1 \quad - \quad 9.52\, b_2 \; + \; 51.00\, b_3 \; = \; -16.36$

In matrix terminology, this set of equations can be written as $\mathbf{Pb} = \mathbf{g}$ or

$$\begin{bmatrix} 213.49 & 30.49 & -32.48 \\ 30.49 & 12.12 & -9.52 \\ -32.48 & -9.52 & 51.00 \end{bmatrix} \begin{bmatrix} b_1 \\ b_2 \\ b_3 \end{bmatrix} = \begin{bmatrix} 87.25 \\ 24.16 \\ -16.36 \end{bmatrix}$$

These three equations can be solved by the method of elimination. This process is based on the principles that any equation can be divided by a constant without changing the equality, and that the difference between two equalities is also an equality. These basic rules are used to eliminate unknown coefficients b_i until only one remains, and then to proceed in a backward fashion to solve for the remaining unknowns. One may proceed as follows.

Eq. IV = Eq. I/213.49: $\quad b_1 + 0.14282\, b_2 - 0.15214\, b_3 = 0.40868$

Eq. V $\;$ = Eq. II/30.49: $\quad b_1 + 0.39751\, b_2 - 0.31223\, b_3 = 0.79239$

Eq. VI = Eq. III/-32.48: $\quad b_1 + 0.29310\, b_2 - 1.57020\, b_3 = 0.50369$

Eq. VII = Eq. V − Eq. IV: $0.25469\ b_2 - 0.16009\ b_3 = 0.38371$

Eq. VIII = Eq. VI − Eq. IV: $0.15028\ b_2 - 1.41806\ b_3 = 0.09501$

Eq. IX = Eq. VII/0.25469: $b_2 - 0.62857\ b_3 = 1.50658$

Eq. X = Eq. VIII/0.15028: $b_2 - 9.43612\ b_3 = 0.63222$

Eq. XI = Eq. IX − Eq. X: $8.80755\ b_3 = 0.87436$

From Equation XI, $b_3 = 0.87436/8.80755 = 0.09927$. Substituting b_3 in Equation IX (or X) gives $b_2 = 0.63222 + 9.43612(0.09927) = 1.56894$. Substituting b_3 and b_2 into Equation IV (or V or VI) gives $b_1 = 0.40868 - 0.14282(1.56894) + 0.15214(0.09927) = 0.19971$. Thus, $b_1 = 0.19971$, $b_2 = 1.56894$, and $b_3 = 0.09927$, and one would use an index $I = 0.1997\ P_A + 1.5689\ P_B + 0.0993\ P_C$ as a basis for selection for improved genotypic values of trait A.

A second method which can be used for calculating the value of the b_i from a set of simultaneous equations is through the application of Cramer's rule. According to Cramer's rule, $b_1 = D_1/D$, $b_2 = D_2/D$, and $b_3 = D_3/D$, where D is the determinant of the matrix **P** and D_1, D_2, and D_3 are the determinants of the matrices formed by replacing columns 1, 2, or 3 of matrix **P** by the column of right hand elements in the set of simultaneous equations. The determinant of

$$\mathbf{P} = \begin{bmatrix} 213.49 & 30.49 & -32.48 \\ 30.49 & 12.12 & -9.52 \\ -32.48 & -9.52 & 51.00 \end{bmatrix}$$

is

$$\begin{aligned} D = &\ (213.49)[(12.12)(51.00) - (9.52)(9.52)] \\ & - (30.49)[(30.49)(51.00) - (9.52)(32.48)] \\ & + (-32.48)[(30.40)(-9.52) - (12.12)(-32.48)] = 71271.7122 \end{aligned}$$

Similarly, the determinant of

$$\mathbf{P}_1 = \begin{bmatrix} 87.25 & 30.49 & -32.48 \\ 24.16 & 12.12 & -9.52 \\ -16.38 & -9.52 & 51.00 \end{bmatrix}$$

is $D_1 = 14231.8390$. Continuing by replacing columns 2 and 3 in turn and finding the determinants of the resulting matrices gives $D_2 = 111802.1416$ and $D_3 = 7042.6642$. From these calculations, the index coefficients can now be calculated as

$$b_1 = 14231.8390/71271.7122 = 0.19968$$

$$b_2 = 111802.1416/71271.7122 = 1.56867$$

$$b_3 = 7042.6642/71271.7122 = 0.09881$$

These values agree, except for rounding errors, with the values determined by the method of elimination.

With access to modern computers, most researchers will opt for the matrix approach to solving simultaneous equations. The vector of unknown index coefficients is solved as $\mathbf{b} = \mathbf{P}^{-1}\mathbf{g}$, where $\mathbf{P}^{-1}$ is the inverse of the phenotypic variance-covariance matrix and $\mathbf{g}$ is the vector of genotypic variances and covariances associated with trait A. For this example, the inverse of the phenotypic matrix is

$$\mathbf{P}^{-1} = \begin{bmatrix} 0.007401 & -0.017479 & 0.001451 \\ -0.017479 & 0.137966 & 0.014622 \\ 0.001451 & 0.014622 & 0.023261 \end{bmatrix}$$

Postmultiplying by $\mathbf{g} = [87.25 \ 24.16 \ -16.38]'$ gives the index coefficients

$$b_1 = 0.007401(87.25) - 0.017479(24.16) + 0.001541(-16.38) = 0.19968$$

$$b_2 = -0.017479(87.25) + 0.137966(24.16) + 0.014622(-16.38) = 1.56867$$

$$b_3 = 0.001451(87.25) + 0.014622(24.16) + 0.023261(-16.38) = 0.09881$$

Again, these estimates are within rounding error of the solutions obtained by the other two methods.

Expected response to selection based on the index, $I = 0.1997 \, P_A + 1.5689 \, P_B + 0.0993 \, P_C$, can be predicted by the equation

$$R_I = i \, \sigma_{G(A)} r_{G(A)I}$$

where i is the standardized selection differential, $\sigma_{G(A)}$ is the genotypic standard deviation for trait A, and $r_{G(A)I}$ is the correlation between the genotypic values for trait A and the index values. Response to selection based on trait A alone is expected to be

$$R_A = i \, \sigma_{G(A)} r_{G(A)P(A)}$$

where $r_{G(A)P(A)}$ is the correlation between the genotypic and phenotypic values of trait A. Note that this correlation equals the square root of the heritability of trait A (i.e., h_A). Thus, direct response to selection for trait A is expected to be equal to $R_A = i \, \sigma_{G(A)} h_A$. To compare expected response from using the index to expected response to direct selection using the same selection intensity, one may calculate

$$\frac{R_I}{R_A} = \frac{r_{G(A)I}}{h_A}$$

The correlation between the genotypic values of trait A and the index can be estimated as

$$r_{G(A)I} = \frac{\sigma_{G(A)I}}{[\sigma_{G(A)}^2 \sigma_I^2]^{0.5}}$$

$$= \frac{b_1\sigma_{G(A)}^2 + b_2\sigma_{G(AB)} + b_3\sigma_{G(AC)}}{[\sigma_{G(A)}^2]^{0.5} [b_1^2\sigma_{P(A)}^2 + 2b_1b_2\sigma_{P(AB)} + 2b_1b_3\sigma_{P(AC)} + b_2^2\sigma_{P(B)}^2 + 2b_2b_3\sigma_{P(AC)} + b_3^2\sigma_{P(C)}^2]^{0.5}}$$

$$= \frac{(0.1997)(87.25) + (1.5689)(24.16) + (0.0993)(-16.38)}{[87.25]^{0.5} [(0.1997)^2(213.49) + \ldots + (0.0993)^2(51.00)]^{0.5}}$$

$$= \frac{53.7109}{(9.3408)(7.3281)} = 0.7845$$

Then, $R_I/R_A = 0.7845/[(87.25)/(213.49)]^{0.5} = 0.7845/0.6392 = 1.23$. This indicates that, for this example, selection based on the index $I = 0.1997\ P_A + 1.5689\ P_B + 0.0993\ P_C$ would be expected to result in 23% greater increase in the mean genotypic value of trait A than would direct selection for trait A at the same selection intensity.

Using matrices, the correlation between genotypic values of trait A and the index values can be expressed as

$$r_{G(A)I} = \frac{\mathbf{b'G}}{[\mathbf{b'Pb}]^{0.5} [\sigma_{G(A)}^2]^{0.5}}$$

where $\mathbf{b'}$ indicates the transpose of the vector $\mathbf{b}$. Using this terminology, the relative efficiency of index selection compared to direct selection can be expressed as

$$\frac{R_I}{R_A} = \frac{[\mathbf{b'G}]\sigma_{P(A)}}{[\mathbf{b'Pb}]^{0.5} \sigma_{G(A)}^2}$$

$$= \frac{(53.7109)(213.49)^{0.5}}{(53.7010)^{0.5} (87.25)} = 1.23$$

Indices can be used to enhance single trait selection for different populations and selection schemes. They may be used to select among partially inbred lines within a cross in a self-pollinated crop, or for mass selection among individuals in a cross-pollinated crop. And they may be used to enhance response to selection for a single trait among families of a particular type. The critical point is to obtain estimates of genotypic and phenotypic variances and covariances for the particular types of selection units and for the traits that are considered to be useful in developing an index. Once such estimates are available, the index coefficients b_i are estimated from the equations $\mathbf{Pb} = \mathbf{g}$ by setting $\mathbf{b} = \mathbf{P}^{-1}\mathbf{g}$. The matrix $\mathbf{P}$ consists of phenotypic variances and covariances of all traits to be included in the index while the vector $\mathbf{g}$ contains the genotypic covariances between all the traits in the index and the one trait of primary interest. If the primary trait is also included in the index, one of the items of $\mathbf{g}$ will be its genotypic variance.

An estimate of the relative efficiency of selection based on an index compared to direct selection for the primary trait itself is given by

$$\frac{R_I}{R_A} = \frac{[\mathbf{b'G}]\ \sigma_{P(A)}}{[\mathbf{b'Pb}]^{0.5}\ \sigma_{G(A)}^2}$$

where the primary trait is designated as trait A, and it is assumed that selection intensity will be the same by both methods of selection.

It is known that the estimate of relative response will be biased upward due to errors in estimation of the population parameters. The amount of bias is unknown in any particular case. However, one should be aware that the estimated efficiency is maximum and that actual response to index selection will not be as great as indicated by the above equation, particularly if estimates of population parameters are not very reliable.

B. Multistage Selection for a Single Trait

Cochran[2] discussed the use of selection indices in multistage selection. The example illustrated by Cochran concerned the evaluation of a number of genotypes in two successive seasons with the objective of selecting the best $1/24^{th}$ of the genotypes. An assumption was that the same amount of resources would be available for testing in each of the two seasons. Thus, if a proportion q_1 of the genotypes were selected in the first season, each of the selected genotypes would be tested in $1/q_1$ as many replications in the second season. The problem was to choose q_1 to maximize overall response to selection.

Consider that the average phenotypic value of the ith genotype in the first season is $P_{i1} = G_i + E_{i1}$, and that the heritability of differences among genotype means in the first season is $h^2 = \sigma_G^2/(\sigma_G^2 + \sigma_{E1}^2)$. Let the ratio of environmental variance to genotypic variance in the first season be $u = \sigma_{E1}^2/\sigma_G^2 = (1 - h^2)/h^2$. Response to selection of a proportion q_1 of the genotypes in the first season is expected to be $R_1 = i_1 \sigma_G h$, where i_1 is the standardized selection differential corresponding to the proportion q_1.

In the second season, each of the selected genotypes will be evaluated in $1/q_1$ as many replications. If the environmental variation is homogeneous over seasons, the environmental variation associated with genotype means in the second season will be $q_1\sigma_{E1}^2 = q_1u\sigma_G^2$, since $u = \sigma_{E1}^2/\sigma_G^2$. Thus, if selection is not practiced in the first season, the phenotypic variance in the second season will be equal to $\sigma_G^2 + q_1u\sigma_G^2 = \sigma_G^2(1 + q_1u)$. Cochran[2] indicated that the regression of the genotypic values, G, on the phenotypic values, P, should remain constant from season to season, even if one selects a portion of the population in the first season. For that reason, the proper weighting of information from the first season with that from the second season can be considered as though selection had not occurred in the first season.

Two types of data will be available as a basis for selection in the second season. The phenotypic values observed in the first season, P_{i1}, will have variance $\sigma_{P1}^2 = \sigma_G^2 + \sigma_{E1}^2 = \sigma_G^2(1 + u)$. The phenotypic values observed in the second season, P_{2i}, will have variance, $\sigma_{P2}^2 = \sigma_G^2(1 + q_1u)$, as shown above. Since the environmental deviations from each of the two years will be uncorrelated, the covariance between the two phenotypic values will be $\sigma_{P1P2} = \sigma_G^2$, the genotypic variance.

From these considerations, the best weighting of the two phenotypic values for selection in the second season is given by the index coefficients derived from solving the following equations.

$$b_1\sigma_G^2(1 + u) + b_2\sigma_G^2 = \sigma_G^2$$

$$b_1\sigma_G^2 + b_2\sigma_G^2(1 + q_1u) = \sigma_G^2$$

The σ_G^2 on the right hand side of each of these two equations represents the covariance between P_{1i} and G_i in the first equation and the covariance between P_{2i} and G_i in the second equation. Solving these two equations gives index coefficients

$$b_1 = q_1u/(q_1u + u), \text{ and } b_2 = u/(q_1u + u)$$

Since only relative values of index coefficients are critical, one may use the relative values $b_1 = q_1$ and $b_2 = 1$. The best index for selection in the second season is therefore given by $I = q_1P_1 + P_2$. As pointed out by Cochran,[2] this is equivalent to utilizing the unweighted mean of all the data which is available on each genotype by the second season.

In calculating expected response to index selection in the second season, it is necessary to recognize the effects of selection in the first season upon the genotypic and phenotypic covariances in the second season. According to Cochran,[2] the genotypic variance in the second season will be

$$\sigma_{G2}^2 = \sigma_G^2 [1 - h^2 i_1 (i_1 - t_1)]$$

where t_1 is the truncation point on the standard normal curve which corresponds to the standardized selection differential i_1. Both i_1 and t_1 can be determined for a particular value of q_1 by reference to a table of the standard normal curve. Cunningham[3] proposed that $i_1(i_1 - t_1)$ be written as s.

Using a generalization given by Cunningham,[3] the phenotypic variances and covariances within the proportion selected in the first season can be estimated. Within the selected group, the variance of phenotypic values observed in the first season is expected to be

$$\sigma_{P1}^2 = \sigma_P^2(1 - s) = \sigma_G^2(1 + u)(1 - s)$$

The variance of the second season phenotypic values should be

$$\sigma_{G1}^2 + \sigma_{E2}^2 = \sigma_G^2(1 - h^2 s) + \sigma_G^2 q_1 u = \sigma_G^2(1 - h^2 s + q_1 u)$$

The covariance between first and second season observations within the selected group is expected to be

$$\sigma_{P1P2}(1 - s) = \sigma_G^2(1 - s)$$

Using this information, expected response to index selection in the second season is given by

$$R_2 = i_2 \frac{\sigma_{G2I}}{\sigma_I} = i_2 \frac{[q_1\sigma_{G2}^2 + \sigma_{G2}^2]}{[q_1^2\sigma_{P1}^2 + 2q_1\sigma_{P1P2} + \sigma_{P2}^2]^{0.5}}$$

$$= i_2 \frac{(q_1 + 1)(1 - h^2 s) \, \sigma_G}{[q_1^2(1 + u)(1 - s) + 2q_1(1 - s) + (1 - h^2 s + q_1 u)]^{0.5}}$$

Thus, total response to selection of a proportion q_1 based on P_1 in the first season, and a proportion q_2 based on the index $I = q_1P_1 + P_2$ in the second season, is $R = R_1 + R_2$ and

$$\frac{R}{\sigma_G} = i_1 h + i_2 \frac{(q_1 + 1)(1 - h^2 s)}{[q_1^2(1 + u)(1 - s) + 2q_1(1 - s) + (1 - h^2 s + q_1 u)]^{0.5}}$$

where h^2 is the heritability in the first season, q_1 and q_2 are the proportions selected in the first and second seasons, and $s = i_1(i_1 - t_1)$ is a function of q_1. Evaluation of this expression for various values of h^2, q_1 and q_2 (Table 1) shows that the total response to selection is usually greatest when $q_1 = q_2 = q^{0.5}$ where q is the overall proportion of genotypes to be selected in the two seasons of testing and selection.

Table 1
EXPECTED RESPONSE TO SELECTION IN TWO STAGES

Proportion selected[a]		Selection intensity			Heritability in stage 1			
Stage 1	Stage 2	i_1	i_2	s	0.7	0.5	0.3	0.1
1	1/24	0.000	2.143	0.000	1.94	1.75	1.46	0.91
1/2	1/12	0.798	1.848	0.636	1.90	1.82	1.62	1.12
1/3	1/8	1.091	1.647	0.721	1.96	1.89	1.70	1.20
1/4	1/6	1.272	1.500	0.759	2.01	1.93	1.74	1.24
1/6	1/4	1.500	1.272	0.802	2.04	1.95	1.75	1.26
1/8	1/3	1.647	1.091	0.818	2.05	1.93	1.72	1.24
1/12	1/2	1.848	0.798	0.865	2.02	1.87	1.63	1.16
1/24	1	2.143	0.000	0.885	1.79	1.52	1.17	0.68

Note: Expected response expressed in genotypic standard deviations.

[a] See text for definitions of q_1, q_2, i_1, i_2, and s.

The formulation given above is equivalent to that of Cochran[2] and applies only to two-stage selection where the same number of plots are to be evaluated in each season. If, instead of testing each genotype in $1/q_1$ as many plots in the second season as in the first, one were to test in $1/p$ as many plots, q_1 should be replaced by p in the above equation and selection in the second season should be based on the index $I = pP_1 + P_2$.

As pointed out by Cochran,[2] the advantage of two stage selection can be great if heritabilities are low to intermediate. With lower heritabilities, discarding some genotypes after one season allows for more critical evaluation of the remaining genotypes in the second season. This can result in overall enhancement of total response to selection.

Cunningham[3] suggested that one might select in the second season just on the basis of the data collected in that season, rather than on the basis of an index incorporating data from both seasons. This approach would avoid the inconvenience of having to assemble the data from both seasons in order to calculate index values. If this was done, response to selection in the second season would be expected to be

$$R_2' = i_2 \frac{\sigma_G^2(1 - h^2 s)}{[1 - h^2 s + q_1 u]^{0.5}}$$

and total response per genotypic standard deviation would be

$$\frac{R'}{\sigma_G} = i_1 h + i_2 \frac{1 - h^2 s}{[1 - h^2 s + q_1 u]^{0.5}}$$

Evaluation of R'/R for the same combinations of parameters considered in Table 1 gave a minimum value of 0.74, a maximum value of 1.00, and an average value of 0.96. It would appear that Cunningham's suggestion[3] will give response sufficiently close to maximum unless the proportion selected in the first season is less than 0.25. If equal resources are applied to both stages of a two-stage selection program, and if the proportion selected in the first stage exceeds about 0.25, it will be quite satisfactory to base selection in the second stage on second stage data only.

Cunningham[3] emphasized that expected response in the second stage will be overestimated because i_2 is calculated from q_2 on the assumption that the phenotypic values in the second stage are normally distributed. Because of the effects of selection in the first stage, this will not be strictly true. However, there will not likely be a serious discrepancy unless the selection intensity in the first stage is high ($q_1 < 0.10$) and/or the heritability of the trait is very high.

Cunningham[3] also mentioned the possibility of selection in a third stage. It is difficult to develop the theory necessary to create an optimum strategy for three-stage selection because of the effects of selection in early stages on the distributions of genotypic values in later stages. From the detailed consideration of two-stage selection given above, one might recommend that proportions selected in each stage of a three-stage selection program be set equal to the cube root of the overall proportion to be selected, and that selection in each stage be based on the data collected in that particular stage.

It is possible to combine two-stage selection with the type of index selection discussed in Section II. A of this chapter, or, for that matter, with any form of index selection. Cunningham[3] has shown how the phenotypic and genotypic variances and covariances should be adjusted after one cycle of index selection at a specified selection intensity. A new index would then be calculated and used for selection in the second stage. This approach can be demonstrated with reference to the simulated data discussed in Section II. A.

In that example, the genotypic variance-covariance matrix was given as

$$\mathbf{G} = \begin{bmatrix} 87.25 & 24.16 & -16.38 \\ 24.16 & 10.26 & -7.15 \\ -16.38 & -7.15 & 30.90 \end{bmatrix}$$

and the phenotypic variance-covariance matrix was given as

$$\mathbf{P} = \begin{bmatrix} 213.49 & 30.49 & -32.48 \\ 30.49 & 12.12 & -9.52 \\ -32.48 & -9.52 & 51.00 \end{bmatrix}$$

Using $\mathbf{b} = \mathbf{P}^{-1}\mathbf{g}$ gave the index $I_1 = 0.1997\ P_A + 1.5689\ P_B + 0.0993\ P_C$ for selection in the first stage (Section II. A). The phenotypic values observed in the first stage for each of the three traits would be incorporated into this index to develop an overall score for each genotype tested at that stage. A proportion, q_1, of the genotypes would then be selected on the basis of the calculated index values.

Once selection has been completed, the genotypic variance-covariance matrix, $\mathbf{G}$, should be adjusted to reflect the effect of selection. Cunningham[3] described the following method for adjusting the matrix. First, construct a matrix in which an extra column and an extra row are added to the matrix $\mathbf{G}$. This extra column and row contain the variance of the index and its covariances with the genotypic values of traits A, B, and C.

The variance of the index

$$I_1 = 0.1997\ P_A + 1.5689\ P_B + 0.0993\ P_C$$

is given by

$$\sigma_{I1}^2 = \mathbf{b}_1'\mathbf{P}\mathbf{b}_1 = 0.1997^2(213.49) + 2(0.1997)(1.5689)(30.49) +$$

$$\dots + (0.0993)^2(51.00) = 53.7010$$

The covariance between the index and the genotypic value of trait A is given by

$$\sigma_{I1G(A)} = 0.1997(87.25) + 1.5689(24.16) + 0.0993(-16.38) = 53.7109$$

Similarly,

$$\sigma_{I1G(B)} = 0.1997(24.16) + 1.5689(10.26) + 0.0993(-7.15) = 20.2117$$

and

$$\sigma_{I1G(C)} = 0.1997(-16.38) + 1.5687(-7.15) + 0.0993(30.90) = -11.4204$$

The figures in brackets are the genotypic covariances between the items in the index and traits A, B, and C. These expressions are based on the assumption that the environmental deviations are not correlated with genotypic values for any trait.

The augmented genotypic covariance matrix becomes

$$\begin{bmatrix} 53.7010 & 53.7109 & 20.2117 & -11.4204 \\ 53.7109 & 87.25 & 24.16 & -16.38 \\ 20.2117 & 24.16 & 10.26 & -7.15 \\ -11.4204 & -16.38 & -7.15 & 30.90 \end{bmatrix}$$

In order to adjust the elements of the genotypic covariance matrix for selection based on I_1, one must assume that the index and the genotypic values of traits A, B, and C are jointly distributed according to a multivariate normal density with a covariance matrix as given above. The amount of adjustment depends upon the selection intensity, q_1, in the following way. With selection of the best q_1 proportion based on the index I_1, the standardized selection differential is i_1 and the corresponding point of truncation of the standard normal curve is t_1. Then, let $s = i_1(i_1 - t_1)$. The adjusted genotypic covariance between traits A and B, say, is then given by

$$\sigma^*_{G(AB)} = \sigma_{G(AB)} - \frac{\sigma_{I1G(a)}\sigma_{I1G(B)}}{\sigma^2_{I1}} s$$

$$= 24.16 - \frac{(53.7109)(20.2117)}{53.7010} s$$

If $q_1 = 0.30$, then $t_1 = 0.525$, $i_1 = 1.182$ and $s = 0.74$. Substituting $s = 0.74$ into the expression above gives $\sigma^*_{G(AB)} = 9.24$. In a similar way,

$$\sigma^{*2}_{G(A)} = 87.25 - (53.7109)(53.7109)(0.74)/(53.7010) = 47.51$$

In matrix notation, the adjusted genotypic covariance matrix is given by

$$\mathbf{G}^* = \mathbf{G} - \frac{[\mathbf{b_i'G}]'[\mathbf{b_i'G}]}{[\mathbf{b_i'Pb_i}]} \, s$$

$$= \begin{bmatrix} 47.51 & 9.24 & -7.93 \\ 9.24 & 4.63 & -3.98 \\ -7.93 & -3.98 & 29.10 \end{bmatrix}$$

if the best 30% are selected on the basis of the index I_1.

Since the genotypes will be retested in the second stage but with, for example, three times as much replication, the environmental covariance matrix should be the same as in the first stage, except that each item will be divided by three. Thus $\mathbf{E}^* = \mathbf{E}/3 = (\mathbf{P} - \mathbf{G})/3$ and the new phenotypic covariance matrix for the second stage should be

$$\mathbf{P}^* = \mathbf{G}^* + (\mathbf{P} - \mathbf{G})/3$$

$$= \begin{bmatrix} 47.51 & 9.24 & -7.93 \\ 9.24 & 4.63 & -3.98 \\ -7.93 & -3.98 & 29.10 \end{bmatrix} + 0.3333 \begin{bmatrix} 126.24 & 6.33 & -16.10 \\ 9.24 & 1.89 & -2.37 \\ -16.10 & -2.37 & 20.10 \end{bmatrix}$$

$$= \begin{bmatrix} 89.59 & 11.35 & -13.30 \\ 11.35 & 5.26 & -4.77 \\ -13.30 & -4.77 & 35.80 \end{bmatrix}$$

Cunningham[3] stated that selection in the first stage would have no effect on the index coefficients calculated for the second stage. In Cunningham's example, data were collected on each individual only once. Some traits were evaluated in the first stage and others in the second stage. In such an instance, whether or not traits measured in the second stage were measured on the total population or only that portion selected on the basis of traits measured in the first stage, the selection has no effect on the index calculated in the second stage. As Cunningham indicated, the multiple regression of overall genotypic worth on the phenotypic values of individual traits should not change when only a portion of the population is studied. Cunningham's conclusion does not apply to the case where data are collected on the same genotypes at two different stages of testing. To see the effect of selection in the first stage on the genotypic parameters of the individuals going into the second stage, certain parameters were compared before and after selection (Table 2). In general, heritabilities are expected to increase because of the additional testing in the second stage. However, genotypic variances as well as genotypic and phenotypic correlations are expected to decrease in absolute magnitude (Table 2).

For selection in the second stage, one would calculate a new index based on the adjusted genotypic and phenotypic covariance matrices. In the example being considered, the new index coefficients would be estimated from the following equations.

Table 2
EFFECTS OF A PARTICULAR INDEX SELECTION ON POPULATION PARAMETERS

Parameter	Before selection	After selection[a]
Genotypic variance — A	87.25	47.51
Genotypic variance — B	10.26	4.63
Genotypic variance — C	30.90	29.10
Heritability — A	0.41	0.53
Heritability — B	0.84	0.88
Heritability — C	0.61	0.81
Genotypic correlation — AB	0.81	0.62
Genotypic correlation — AC	−0.32	−0.21
Genotypic correlation — BC	−0.40	−0.38

[a] Selection of the best 30% based on the index $I = 0.1997\ P_A + 1.5689\ P_B + 0.0993\ P_C$. Variances adjusted on the assumption of normality. Heritabilities after selection reflect three-fold increase in replication (see text).

$$89.59\ b_A + 11.35\ b_B - 13.30\ b_C = 47.51$$

$$11.35\ b_A + 5.26\ b_B - 4.77\ b_C = 9.24$$

$$13.30\ b_A - 4.77\ b_B + 35.80\ b_C = -7.93$$

Solving these equations gives the new index

$$I_2 = 0.4261\ P_A + 0.8871\ P_B + 0.0550\ P_C$$

which should be used as the basis for selection in the second stage. Since the index used in the second stage differs from that used in the first stage, it is doubtful that the optimum selection in the second stage should be based on the overall index $I = I_1 + 3I_2$, as would be the case if one were selecting for a single trait. However, selection for I_1 in the first stage and I_2 in the second stage should give response sufficiently close to the optimum to make this a reasonable approach to the problem of two-stage selection with an index.

The above discussion of two-stage selection was based on the assumption that environmental variation and covariation was homogeneous over stages. This may not be true in practice. However, such heterogeneity should not change the overall strategy of selecting with approximately equal intensity in each stage and of selecting in each stage on the basis of only the data gathered in that stage. The presence of variation due to interaction between genotypes and seasons and/or locations has the effect of increasing the overall environmental variation and increasing the advantages of two-stage selection over single stage selection.

A possible strategy for two-stage selection is to select perhaps one-fifth to one-quarter of the genotypes on the basis of a single replicated test at one location in the first stage, and to test the selected genotypes with the same amount of replication, but at four or five locations in the second stage. This is comparable to having both stages conducted at one location where each genotype is replicated four to five times as often in the second stage as in the first. However, genotype-environment interaction would now be included as part of the environmental variation rather than as part of the genotypic variation at a particular location. If this type of interaction is important, multilocation testing in the second stage will give greater response in true genotypic value than would testing at a single location.

C. Information from Relatives

Falconer[4] described the merits of weighting data from different families and individuals when selecting within populations in which there is some type of family structure. Depending upon the phenotypic and genotypic correlations among individuals within families, selection based on an index including family means and individual values will be as efficient, or more efficient, than selection based on individual performance, family performance, or within-family deviations.

The method can be described in a general case by considering a population composed of families each containing an equal number of individuals. Define $P_f = G_f + E_f$ as the phenotypic mean of a particular family and $P_w = G_w + E_w$ as the phenotypic value of an individual expressed as a deviation from the mean of its respective family. In index terminology, the objective is to improve genotypic worth ($W = G_f + G_w$) by using an index $I = b_f P_f + b_w P_w$. The index coefficients, b_f and b_w, can be calculated from the following pair of simultaneous equations.

$$b_f \, \sigma_{P(f)}^2 + b_w \, \sigma_{P(fw)} = \sigma_{G(f)}^2 + \sigma_{G(fw)}$$

$$b_f \, \sigma_{P(fw)} + b_w \, \sigma_{P(w)}^2 = \sigma_{G(fw)} + \sigma_{G(w)}^2$$

However, since the within-family deviations are not correlated with the family means, the equations reduce to

$$b_f \, \sigma_{P(f)}^2 = \sigma_{G(f)}^2 \text{ , and } b_w \, \sigma_{P(w)}^2 = \sigma_{G(w)}^2$$

with solutions

$$b_f = \sigma_{G(f)}^2 / \sigma_{P(f)}^2 = h_f^2 \text{ , and } b_w = \sigma_{G(w)}^2 / \sigma_{P(w)}^2 = h_w^2$$

As indicated by Falconer,[4] the optimum improvement in genotypic value should result from selecting on the basis of an index in which family means and within-family deviations are each weighted according to their respective heritabilities.

If h^2 is the heritability of individual observations in the population, Falconer[4] showed that

$$h_f^2 = h^2 \, \frac{1 + (n - 1)r}{1 + (n - 1)t}$$

and

$$h_w^2 = h^2 \, \frac{1 - r}{1 - t}$$

where r and t are the genotypic and phenotypic intraclass correlations for the particular type of family being considered, and n is the number of individuals in each family. The genotypic intraclass correlation, r, is expected to be 0.5 for full-sib families and 0.25 for half-sib families in a random mating population, provided that the trait under consideration is controlled primarily by additive genes.

The phenotypic intraclass correlation, t, depends in part on the degree of genetic relationship between individuals within a family and in part on the extent to which individuals within the same family are subject to common environmental effects. In plant populations, proper randomization can remove environmental effects which cause phenotypic similarity between individuals within the same family. As long as individuals are not grouped according to their family structure, the phenotypic correlation between individuals within the same family should be equal to the phenotypic variance among family means divided by the total phenotypic variance among individuals within the population. That is, $t = r\sigma_G^2 + \sigma_E^2/n)/(\sigma_G^2 + \sigma_E^2)$. For families of large size, $t = rh^2$.

Thus, if families are properly randomized and of size n, the optimum selection rule is to select on the basis of the index $I = h_f^2 P_f + h_w^2 P_w$, where

$$h_f^2 = h^2 \frac{1 + (n - 1)r}{1 + (n - 1)rh^2}$$

$$h_w^2 = h^2 \frac{1 - r}{1 - rh^2}$$

Expected response to this combined selection is given by

$$R_C = i\,\sigma_G\,h\left[1 + \frac{r^2(1 - h^2)(n - 1)}{1 + (n - 1)rh^2}\right]^{0.5}$$

Similarly, response to individual selection is $R_I = i\sigma_G h$.

The efficiency of combined selection compared to individual selection at the same selection intensity can be assessed by calculating

$$\frac{R_C}{R_I} = \left[1 + \frac{r^2(1 - h^2)(n - 1)}{1 + (n - 1)rh^2}\right]^{0.5}$$

where h^2 is the heritability of observations on individuals, n is the number of individuals in each family, and r is the average genotypic correlation between individuals in the same family. If one was selecting within an inbred population with family structure, h^2 could be interpreted as the heritability of the mean observation on each line, n as the number of lines per family, and r as the genotypic correlation between line means within families.

The relative efficiency of combined selection compared to individual selection is greatest when heritability is low, when there are a large number of individuals per family, and when there is a high genotypic correlation between individuals within families (Table 3). The genotypic correlation between individuals within a family is expected to be 0.25 for half-sib families and 0.5 for full-sib families in a random mating population. The genotypic correlation may exceed 0.5 in partially inbred populations. For example, if one is selecting among F_4-derived lines which can be grouped together in families derived from F_3 parents, the expected genotypic correlation between lines within the same family is 0.86. In such a case, combined selection would increase expected response to selection by at least 20% if each family were represented by eight or more F_4-derived lines and if heritability of line means was 0.5 or less (Table 3).

The technique of combining data from individual observations with data on family means seems to be more promising for improving response to selection for single traits in partially inbred populations. The one condition is that lines have been properly randomized when tested. If grouped by cross or by family, the phenotypic correlation between individual lines within families will increase to such an extent that the advantages of combined selection will be minimal.

To apply this method to a self-pollinated crop, one would require an estimate of heritability of line means, h^2, and a theoretical estimate of r, the genotypic correlation between line

Table 3
RELATIVE EFFICIENCY OF COMBINED
INDIVIDUAL AND FAMILY SELECTION TO
INDIVIDUAL SELECTION WHEN THERE ARE NO
COMMON ENVIRONMENTAL EFFECTS ON
INDIVIDUALS WITHIN THE SAME FAMILY

Heritability	Genotypic correlation	Relative efficiency (%) for family size of:			
		2	4	8	16
0.2	0.25	102	105	110	117
0.5	0.25	101	102	103	105
0.8	0.25	100	100	100	101
0.2	0.50	108	119	132	144
0.5	0.50	103	107	110	112
0.8	0.50	101	101	102	102
0.2	0.75	117	137	156	172
0.5	0.75	107	115	120	123
0.8	0.75	102	103	104	104

means within the same family. Estimates for these two sources of information, along with the number of lines per family, would allow one to calculate the heritabilities of family means and of within-family deviations. The most useful way of expressing the desired index is as

$$I = P + \frac{rn(1 - h^2)}{(1 - r)[1 + (n - 1)rh^2]} P_f$$

where P is the observation on an individual line, and P_f is the mean of the family in which it occurs. Such an index would be calculated for each individual within each family and the individuals with the best index scores would be selected.

Similar index techniques can be developed for cross-pollinated crops. Moreno-Gonzalez and Hallauer[5] recommended this technique to enhance the effectiveness of full-sib reciprocal recurrent selection in maize, *Zea mays*. In their proposal, genotypic worth was defined as the breeding value of an individual in one or the other of two populations. In each case, selection was to be based on information on full-sib families obtained by crossing one parent from one population with another from the other population, as well as on S_2 families derived by selfing each of the two parents. Two indices would be calculated; one for improving population A and one for improving population B. Moreno-Gonzalez and Hallauer showed that the advantage of using an index approach would be substantial if heritabilities were as low as is typical for grain yield in maize.

III. MULTIPLE TRAIT SELECTION

A. Weight-Free Index for Ranking Genotypes

When it is difficult to specify relative economic values of a number of traits, and when little is known about genotypic and phenotypic variances and covariances within the population, the weight-free index developed by Elston[6] can provide an objective rule for selection. The method ranks as best those individuals which are desirable in all traits. However, the method does allow strengths in one trait to compensate for weaknesses in others.

The method can be described by reference to the simulated data given in Table 4. In this

Table 4
DATA FOR DEMONSTRATING WEIGHT-FREE INDEX CALCULATIONS

Genotype	Phenotypic value (P)				Logarithm (P − k)			Index	Rank
	A	B	C	1/B	A	1/Bᵃ	Cᵃ		
1	92	19.0	6	0.053	2.80	2.69	1.79	13.0	19
2	106	17.2	11	0.058	3.41	2.99	4.03	41.1	2
3	110	23.3	8	0.043	3.54	1.57	3.26	18.1	13
4	84	16.6	7	0.060	2.13	3.08	2.77	18.2	12
5	97	17.9	10	0.056	3.06	2.88	3.83	33.8	3
6	101	13.5	7	0.074	3.24	3.58	2.77	32.1	7
7	101	18.6	8	0.054	3.23	2.76	3.26	29.1	9
8	95	15.1	6	0.066	2.97	3.33	1.79	17.7	14
9	116	18.4	8	0.054	3.70	2.76	3.26	33.3	5.5
10	90	19.9	10	0.050	2.67	2.47	3.83	25.3	10
11	78	14.0	16	0.071	0.88	3.49	4.66	14.3	17
12	107	17.3	12	0.058	3.45	2.99	4.19	43.2	1
13	109	21.1	6	0.047	3.51	2.17	1.79	13.6	18
14	98	24.8	11	0.040	3.11	0.59	4.03	7.4	20
15	114	20.6	10	0.049	3.65	2.38	3.83	33.3	5.5
16	97	23.1	8	0.043	3.06	1.57	3.26	15.7	15
17	116	22.8	9	0.044	3.70	1.76	3.58	23.3	11
18	90	19.3	17	0.052	2.67	2.63	4.75	33.4	4
19	94	23.2	8	0.043	2.91	1.57	3.26	14.9	16
20	91	16.2	9	0.062	2.73	3.17	3.58	31.0	8
Min	78	13.5	6	0.040					
Max	116	24.8	17	0.074					
kᵇ	75.6		5.4	0.0382					

ᵃ Data for trait 1/B multiplied by 1000 and trait C multiplied by 10 before taking logarithms.
ᵇ $k = [n(\text{minimum}) - \text{maximum}]/(n-1) = [20(\text{minimum}) - \text{maximum}]/19$.

example, high values of traits A and C are preferred, while low values of trait B are best. The first step in developing an index is to recode the values for trait B so that preferred genotypes have high values. This can be done by changing the sign for each value of trait B, or by taking the reciprocal of each value. The latter approach has been adopted in this example.

Elston[6] recommended that one compare the histograms of each trait to see that the distributions of the various traits are similar, at least to the extent of having the same number of modes. Because of the small sample size, histograms of the data from Table 4 show little evidence of differences in distribution, although trait C shows some skewness towards higher values. To make the distributions more comparable, Elston[6] proposed use of the transformation $P' = \ln(P - k)$ where $k = [n(\text{minimum value}) - (\text{maximum value})]/(n - 1)$ is chosen so that the distributions will have similar locations and all values will be positive and nonzero. In order to avoid negative logarithms, the reciprocal of trait B was multiplied by 1000 and the values for trait C were multiplied by 10 before applying the transformation to each of the three variables in Table 4. After transformation, the transformed variables had minimum values of 0.88, 0.59, and 1.79, and maximum values of 3.70, 3.58, and 4.75; all reasonably close to each other. The final index for ranking the 20 genotypes was calculated as the product of the transformed scores for each trait.

This method provides a procedure for selecting for several traits when little is known except that high or low values of each trait are desired in the selected material. In effect, each of the traits used in the ranking procedure is weighted approximately equally in terms of their standard deviations.

B. Base Indices

If relative economic values can be specified for each trait, but if reliable estimates of genotypic and phenotypic parameters are not available, the base index suggested by Williams[7] can be used for the simultaneous improvement of two or more traits. In this case, an index is calculated for each genotype by weighting the observed phenotypic values of each trait by its respective economic value and summing over all traits with nonzero economic weight; i.e., $I = a_1P_1 + a_2P_2 + \ldots + a_nP_n$

An example of this type of index was described by Eagles and Frey.[8] They considered that grain yield of oat, *Avena sativa*, was twice as valuable as straw yield. Relative economic values were therefore set at 1.0 for grain yield and 0.5 for straw yield. The base index for improving both grain yield and straw yield would be I = grain yield + 0.5 straw yield. Eagles and Frey calculated a base index score for each of 1200 F_9-derived oat lines at each of three locations in Iowa, as well as for the average values over the three locations. The best lines were selected according to their base index scores. Progeny of the selected and the unselected lines were evaluated at the same locations in subsequent years to assess the response to selection based on the base index as well as four other selection methods.

C. Modified Base Index

Smith et al.[9] recommended that multiple trait selection be based on an index in which the phenotypic value of each trait was weighted according to its heritability rather than according to its economic value. They concluded that an index based on heritability estimates would be more effective than a base index when relative economic values were the same for all traits. When economic values are not of the same magnitude, and when prior knowledge suggests that the heritabilities differ greatly for the traits which are to be improved, it seems reasonable to combine the suggestions of Smith et al.[9] and Williams[7] and use an index in which each trait is weighted according to the product of its heritability and its relative economic value. This proposal follows directly from the derivation of the optimum index developed by Henderson[10] in those cases where traits are independent and selection is based only on those traits which are economically important (see Chapter 7).

If genotypic and phenotypic parameters are not known with sufficient precision, or if correlations are known to be small in absolute magnitude, and if heritabilities and relative economic values are known to vary for important traits, this modified index can provide a useful basis for selection. For each trait, one would calculate $w_i = a_ih_i^2$, where a_i is the relative economic value and h_i^2 is the heritability. Then, for each genotype, one would calculate the index scores from $I = w_1P_1 + w_2P_2 + \ldots + w_nP_n$. These calculated index scores would provide the basis for the simultaneous selection and improvement of all traits included in the index.

In the example described by Smith et al.,[9] relative economic values of grain yield, percentage grain moisture, root lodging, and stalk lodging were 1, -1, -1, and -1, respectively. Over seven populations, average heritabilities were 0.66, 0.74, 0.58, and 0.55 for the four traits. Based on average estimates of heritabilities, one would calculate $w_1 = (1)(0.66) = 0.66$, $w_2 = (-1)(0.74) = -0.74$, $w_3 = (-1)(0.58) = -0.58$, and $w_4 = (-1)(0.55) = -0.55$. Genotypes would be selected on the basis of the index scores calculated as I = 0.66(grain yield) -0.74 (grain moisture) $- 0.58$ (root lodging) $- 0.55$ (stalk lodging).

The extension of this approach to other situations should be straightforward provided that relative economic values are known and that one has estimates of heritabilities of the economically important traits. In some traits, such as lodging, heritability may vary from season to season or from site to site. In such cases, it should be easy to alter this index to reflect those changes. It should be clear that neither the base index, nor this modification of it, allow the use of a correlated secondary trait to enhance response to selection for economically important primary traits.

D. Optimum Selection Indices

The approach described by Henderson[10] is recommended for developing optimum selection indices. Primary traits are considered to be those which have nonzero relative economic values in the breeding program. Secondary traits are those which have zero economic value, but which may be related to primary traits in such a way as to be useful in the selection program. For grain yield of cereal crops, data are often collected on grain yield and one or more of its three components; viz., heads per unit area, kernels per head, and kernel weight. Unless kernel weight is an important trait in grain or seed quality, the three components should be considered as secondary traits and economic importance should be attached only to the primary trait, grain yield.

Henderson's approach[10] to developing an optimum index is to use the methods described in Section II.A of this chapter to develop an index for each primary trait. For this part of the method, there is no need to have estimates of relative economic value. One needs only to know which traits are to be considered as primary traits. It is appropriate to consider other traits which are economically important in their own right as being secondary traits for the improvement of any particular primary trait. Thus, in developing an index for improved grain yield, it may be valuable to include protein concentration as a secondary trait even if another index will be developed for the purpose of improving protein concentration.

The methods described in Section II.A would be used to develop an index for each of, say, m economically important traits. Thus,

$$I_1 = b_{11}P_1 + b_{12}P_2 + \ldots + b_{1n}P_n$$

$$I_2 = b_{21}P_1 + b_{22}P_2 + \ldots + b_{2n}P_n$$

$$\cdot \quad \cdot \quad \cdot \quad \cdot \quad \cdot$$

$$I_m = b_{m1}P_1 + b_{m2}P_2 + \ldots + b_{mn}P_n$$

Note that any b_{ij} may be zero if the jth trait does not contribute to the improvement of the ith primary trait. The final index for the simultaneous improvement of all m economically important traits would then be calculated as

$$I = a_1I_1 + a_2I_2 + \ldots + a_mI_m$$

$$= a_1b_{11}P_1 + a_1b_{12}P_2 + \ldots + a_1b_{1n}P_n$$

$$+ a_2b_{21}P_1 + a_2b_{22}P_2 + \ldots + a_2b_{2n}P_n$$

$$\cdot \quad \cdot \quad \cdot \quad \cdot \quad \cdot \quad \cdot$$

$$+ a_mb_{m1}P_1 + a_mb_{m2}P_2 + \ldots + a_mb_{mn}P_n$$

$$= (a_1b_{11} + a_2b_{21} + \ldots + a_mb_{m1})P_1$$

$$+ (a_1b_{12} + a_2b_{22} + \ldots + a_mb_{m2})P_2$$

$$\cdot \quad \cdot \quad \cdot \quad \cdot \quad \cdot$$

$$+ (a_1b_{1n} + a_2b_{2n} + \ldots + a_mb_{mn})P_n$$

Selection among genotypes would be based on the index scores given by this final index equation.

The advantage of this method lies in the ability to calculate a new index, I', by giving new weights to the individual indices if the relative economic values of the primary traits change. This can be done without recourse to recalculation of the individual indices. To demonstrate this method, the multivariate simulation program used in Section II.A was used to generate a second set of variates labelled as traits D and E. In this case, an index, $I = b_D P_D + b_E P_E$, was calculated to maximize genotypic values in trait D. For this example, it is assumed that none of the traits A, B, or C is useful for improving trait D, and that neither trait D nor E is useful for improving trait A.

For traits D and E, the genotypic and phenotypic covariance matrices for a sample of 200 were

$$\mathbf{G} = \begin{bmatrix} 55.9 & -12.8 \\ -12.8 & 9.36 \end{bmatrix}$$

and

$$\mathbf{P} = \begin{bmatrix} 162.9 & -11.9 \\ -11.9 & 9.58 \end{bmatrix}$$

This gives rise to the following set of simultaneous equations to be solved for index coefficients b_D and b_E.

$$162.9\, b_D - 11.9\, b_E = 55.9$$

$$-11.9\, b_D + 9.58\, b_E = -12.8$$

Solving these equations gives $b_D = 0.27$ and $b_E = -1.00$, and the resulting index is $I_D = 0.27\, P_D - 1.00\, P_E$

The index estimated to be most efficient for improving trait A (Section II. A) was $I_A = 0.20\, P_A + 1.57\, P_B + 0.10\, P_C$. Thus, if the primary traits A and D have relative economic values of 1.0 and 0.3, the best index for improving the two traits would be

$$I = I_A + 0.3\, I_D$$

$$= 0.20\, P_A + 1.57\, P_B + 0.10\, P_C + 0.08\, P_D - 0.30\, P_E$$

If, for some reason, it was decided that more realistic economic values were 1.0 and 1.5 for traits A and D, the new index could be calculated simply as

$$I' = I_A + 1.5\, I_D$$

$$= 0.20\, P_A + 1.57\, P_B + 0.10\, P_C + 0.40\, P_D - 1.50\, P_E$$

E. Restricted Selection Indices

When reliable estimates of phenotypic and genotypic variances and covariances are available, it is possible to develop an index which is expected to give responses in certain traits proportional to specified desired changes. No restrictions would be required for secondary traits in this type of problem. As explained in Chapter 7, this method was proposed independently by Harville[11] in 1975 and by Tai[12] in 1977. The method is a modification of a general formulation first proposed by Tallis.[13] If the index is to include only primary traits for which restrictions are to be specified, the method is equivalent to that of Pesek and Baker.[14]

Suppose that a total of m traits are to be included in the index and that it is required that r of these (r < m) be changed by amounts k_i, i = 1, 2, . . . r. If **P** is the matrix of phenotypic covariances among the m traits, G_r is the r × m matrix of genotypic covariances between the r restricted traits and all m traits, and **k** is an r × 1 vector of desired changes in the restricted traits, the m index coefficients can be estimated from

$$\mathbf{b} = \mathbf{P}^{-1}\mathbf{G}_r'(\mathbf{G}_r\mathbf{P}^{-1}\mathbf{G}_r')^{-1}\mathbf{k} \tag{4}$$

Selection based on the index

$$I = b_1 P_1 + b_2 P_2 + \ldots b_m P_m$$

is expected to give responses in the r restricted traits which are proportional to the specified desired changes. This method can be described in more detail by reference to the data given by Tai.[12]

Consider four traits including marketable yield, chipping score, boiling score, and baking score of potato, *Solanum tuberosum*, tubers. Tai[12] gave estimates of the genotypic and phenotypic covariance matrices for these four traits in 16 potato lines as

$$\mathbf{G} = \begin{bmatrix} 109.23 & -96.23 & 20.59 & 12.50 \\ -96.23 & 172.26 & -26.07 & -10.22 \\ 20.59 & -26.07 & 23.11 & 18.35 \\ 12.50 & -10.22 & 18.35 & 19.49 \end{bmatrix}$$

and

$$\mathbf{P} = \begin{bmatrix} 138.33 & -97.89 & 18.56 & 12.93 \\ -97.89 & 195.74 & -25.07 & -8.59 \\ 18.56 & -25.07 & 32.37 & 20.61 \\ 12.93 & -8.59 & 20.61 & 25.08 \end{bmatrix}$$

In the example considered by Tai, restrictions were placed on the first two traits. Desired change in marketable yield was set at 9.08 kg/plot, while that for chipping score was set at 10 points. In the terminology given above, m = 4, r = 2, k_1 = 9.08, k_2 = 10, and **G** and **P** are as given above. Matrix algebra can be used to solve for the index coefficients **b** = $[b_1 \ b_2 \ b_3 \ b_4]'$ according to Equation 4. The results are b_1 = 0.2616, b_2 = 0.2080, b_3 = 0.0247, and b_4 = 0.0008, which are identical to those calculated by Tai using a different procedure.

Tai's approach[12] was equivalent to setting up a total of 2m equations in 2m unknowns consisting of m (= 4 in the case) index coefficients, b_i, m − r expected gains in nonrestricted traits, and r relative economic values in the restricted traits. The Harville[11] approach (Equation 4) uses the method of Lagrangian multipliers to incorporate the necessary restrictions into m equations. Both methods give the same results.

Expected response to selection based on the index $I = \mathbf{b'P}$ for any particular trait, say the jth, is given by

$$i \, \sigma_I \, b_{G(j)I} = i \, \sigma_I \, \frac{\sigma_{G(j)I}}{\sigma_I^2} = i \, \frac{\sigma_{G(j)I}}{\sigma_I}$$

Since i/σ_I is constant for all traits, the relative responses in each of the traits can be studied by comparing the covariance between their genotypic values and the selection index. It is useful to do that for the example above to see the expected responses to the restricted index, $I = 0.2616 \, P_1 + 0.2080 \, P_2 + 0.0247 \, P_3 + 0.0008 \, P_4$. For the first trait, marketable yield, the covariance between genotypic value and the selection index is estimated by

$$(0.2616)(109.23) + (0.2080)(-96.23) + (0.0247)(20.59) + (0.0008)(12.50) = 9.08$$

The covariance between the genotypic value for the second trait, chipping score, and the selection index is estimated as

$$(0.2616)(-96.23) + (0.2080)(195.74) + (0.0247)(-26.07) + (0.0008)(-10.22) = 10.00$$

Similar calculations give relative expected responses of 0.55 for boiling score and 1.61 for baking score. Note that the relative responses for the restricted traits, marketable yield and chipping score, are proportional to the specified desired changes for those two traits. In actual applications, the ratio of the standardized selection differential, i, to the standard deviation of the index, σ_I, will be less than unity, and the expected responses will be proportional to, but less than, the specified desired responses.

IV. SUMMARY

Selection index methods that are considered to be useful in crop improvement programs have been reviewed, and the methods required for their application have been demonstrated by use of simulated or published data. The methods include the use of secondary traits, multistage selection, and information from relatives to enhance response to selection for improvement of a single economically important quantitative trait.

For crop improvement programs where the simultaneous improvement of several traits is the objective, the methods of weight-free ranking, a base index, a modified base index, and the estimated optimal index have been described and illustrated. An effort has been made to indicate which conditions require which of the choices of indices. In addition, the methodology of restricted indices is demonstrated by reference to published data.

The methods described in this chapter should provide a sound basis for applying the most useful selection indices to a wide range of crop improvement problems.

REFERENCES

1. **Smith, H. F.**, A discriminant function for plant selection, *Ann. Eugenics*, 7, 240, 1936.
2. **Cochran, W. G.**, Improvement by means of selection, Proc. 2nd Berkeley Symp. Math. Stat. Prob., 1951, 449.
3. **Cunningham, E. P.**, Multi-stage index selection, *Theor. Appl. Genet.*, 46, 55, 1975.
4. **Falconer, D. S.**, *Introduction to Quantitative Genetics*, 2nd ed., Longman, New York, 1981.
5. **Moreno-Gonzalez, J. and Hallauer, A. R.**, Combined S_2 and crossbred family selection in full-sib reciprocal recurrent selection, *Theor. Appl. Genet.*, 61, 353, 1982.

6. **Elston, R. C.,** A weight-free index for the purpose of ranking or selection with respect to several traits at a time, *Biometrics,* 19, 85, 1963.
7. **Williams, J. S.,** The evaluation of a selection index, *Biometrics,* 18, 375, 1962.
8. **Eagles, H. A. and Frey, K. J.,** Expected and actual gains in economic value of oat lines from five selection methods, *Crop Sci.,* 14, 861, 1974.
9. **Smith, O. S., Hallauer, A. R., and Russell, W. A.,** Use of index selection in recurrent selection programs in maize, *Euphytica,* 30, 611, 1981.
10. **Henderson, C. R.,** Selection index and expected genetic advance, in Statistical Genetics and Plant Breeding, No. 982, Hanson, W. D. and Robinson, H. F., Eds., National Academy of Sciences, National Research Council, Washington, D.C., 1963, 141.
11. **Harville, D. A.,** Index selection with proportionality constraints, *Biometrics,* 31, 223, 1975.
12. **Tai, G. C. C.,** Index selection with desired gains, *Crop Sci.,* 17, 182, 1977.
13. **Tallis, G. M.,** A selection index for optimum genotype, *Biometrics,* 18, 120, 1962.
14. **Pesek, J. and Baker, R. J.,** Desired improvement in relation to selection indices, *Can. J. Plant Sci.,* 49, 803, 1969.

Chapter 9

PARENTAL EVALUATION

I. INTRODUCTION

Evaluation and choice of parents is an important part of any crop improvement program. In most crops, parental evaluation involves the assessment of several traits which may differ in relative importance. These differences need to be considered when choosing parents for a breeding program.

The choice of parents will depend upon the goals of the breeding program. If the choice is to be made objectively, it will be necessary to develop well-defined goals. These goals may be specified in terms of how much each trait is to be altered from current levels in order to meet the objectives of the program, or by specifying an ideal genotype toward which the breeding program is directed.

Applied plant breeding can be categorized as having short-term or long-term objectives. Plant breeders must recognize that one set of parents may be best for meeting short-term objectives, while another will be best for meeting long-term objectives. For short-term objectives, it seems best to choose parents that will produce a population whose mean values are as close as possible to those specified for the corresponding traits of an ideal genotype. For long-term objectives, on the other hand, it is important to have sufficient genotypic variability to maintain continued response over several cycles of selection. This is particularly true for the economically important traits in the breeding program.

The purpose of this chapter is to describe and illustrate methods for (1) assigning relative importance to different traits in a breeding program, (2) developing objectives in terms of an ideal genotype, and (3) choosing parents to develop populations with suitable mean values or sufficient genotypic variability. Simulated data will be used to illustrate various ideas and techniques.

II. METHODS FOR ASSIGNING RELATIVE IMPORTANCE TO TRAITS IN A BREEDING PROGRAM

A. Proposal by Grafius

Grafius[1] proposed that relative importance of various traits could be assessed by relating the data for those traits to an overall excellence score. If a sample of genotypes is evaluated in a replicated experiment at a particular site, and if a knowledgeable individual was to give an overall rating to those genotypes, Grafius argued that multiple regression techniques could then be used to determine the relative importance of the traits. An important part of this proposal is its reliance on the subjective judgment of someone who is familiar with the crop and who is knowledgeable of what is required in improved cultivars. The overall score for genotypes must be based on data from economically important traits only, and must reflect the importance that the breeder attaches to those traits.

The relative importance of individual traits is obtained by carrying out a multiple linear regression of overall scores on the data available for each trait. The regression coefficients are standardized by multiplying each regression coefficient by the standard deviation of the overall worth score, and by dividing by the standard deviation of the corresponding trait. This standardized partial regression coefficient represents the relative importance of the trait in determining overall worth as judged by a knowledgeable plant breeder.

The process is demonstrated by reference to the simulated data presented in Table 1. These data can be considered as average data for each of five important traits, where averages have been taken over several replications in each of perhaps several environments.

Table 1
SIMULATED VALUES FOR FIVE TRAITS AND OVERALL WORTH OF EACH OF 20 POTENTIAL PARENTS

Parent	A	B	C	D	E	Overall worth
1	31.7	28.1	401	43.2	104.8	32
2	26.4	44.0	453	48.0	104.8	72
3	28.2	39.4	387	55.4	100.7	40
4	28.9	44.4	396	56.3	90.9	68
5	28.4	48.1	440	41.7	104.2	76
6	28.9	42.3	406	50.3	109.3	40
7	34.3	46.6	391	49.7	90.6	75
8	30.7	45.5	428	49.7	94.1	82
9	34.9	41.3	427	57.2	98.5	67
10	39.8	31.6	367	47.4	107.2	19
11	31.2	47.7	395	40.0	95.4	72
12	29.9	44.5	402	59.1	96.4	60
13	31.2	37.6	411	62.1	96.8	54
14	25.9	47.2	419	49.2	94.8	77
15	26.9	44.1	396	43.2	104.5	49
16	28.6	49.2	388	45.2	106.9	47
17	27.9	32.1	392	37.5	99.7	42
18	33.7	41.2	420	42.5	99.1	69
19	20.3	47.9	439	48.1	104.6	68
20	34.0	34.9	429	52.1	106.8	47
Mean	30.1	41.9	409	48.9	100.5	58
s.d.	5.5	6.5	20	7.0	10.0	7.6
b_i	0.11	0.45	0.57	−0.18	−0.57	

Partial regression coefficients can be calculated by using any standard program for multiple linear regression, and they can then be standardized as indicated above. Another way of achieving the same objective is to calculate the correlations among each of the five traits, as well as those between each of the traits and the overall score. These correlations can be inserted into the following set of simultaneous equations, and the equations can be solved to give the required estimates of relative importance of the various traits.

$$b_1 + r_{12}b_2 + r_{13}b_3 + r_{14}b_4 + r_{15}b_5 = r_{1s}$$

$$r_{12}b_1 + b_2 + r_{23}b_3 + r_{24}b_4 + r_{25}b_5 = r_{2s}$$

$$r_{13}b_1 + r_{23}b_2 + b_3 + r_{34}b_4 + r_{35}b_5 = r_{3s}$$

$$r_{14}b_1 + r_{24}b_2 + r_{34}b_3 + b_4 + r_{45}b_5 = r_{4s}$$

$$r_{15}b_1 + r_{25}b_2 + r_{35}b_3 + r_{45}b_4 + b_5 = r_{5s}$$

In this set of equations, r_{ij} refers to the correlation between traits i and j, r_{is} refers to the correlation between the ith trait and the overall score, and the b_i terms represent the standardized partial regression coefficients. For the data given in Table 1, the following equations were developed

$$b_1 - 0.471\,b_2 - 0.393\,b_3 + 0.116\,b_4 - 0.070\,b_5 = -0.310$$

$$-0.471\,b_1 + b_2 + 0.308\,b_3 + 0.027\,b_4 - 0.295\,b_5 = 0.743$$

$$-0.393\,b_1 + 0.308\,b_2 + b_3 + 0.054\,b_4 + 0.084\,b_5 = 0.613$$

$$0.116\ b_1 + 0.027\ b_2 + 0.054\ b_3 + b_4 - 0.293\ b_5 = 0.046$$

$$-0.070\ b_1 - 0.295\ b_2 + 0.084\ b_3 - 0.293\ b_4 + b_5 = -0.615$$

The solutions to these five equations were found to be $b_1 = 0.11$, $b_2 = 0.45$, $b_3 = 0.57$, $b_4 = -0.18$, and $b_5 = -0.57$. Thus, for this set of data, and the assigned subjective scores, the relative importance of traits A, B, C, D, and E is given by the coefficients 0.11, 0.45, 0.57, −0.18, and −0.57, respectively. Traits C and E are judged to be of the same relative importance except that smaller values of trait E are preferred. The relative importances relate to increases of one standard deviation in each trait. For example, a change of one standard deviation in trait A is judged to contribute 0.11 standard deviations to the overall score. These weights would be the weights used in subsequent evaluation of parental material and choice of parents for improving the overall score in this hypothetical crop.

Assigning overall scores to a sample of genotypes is subjective and will be highly dependent upon the knowledge and experience of the person doing the scoring. In actual practice, one would probably compare the data for each trait with the data for a standard commercial cultivar and develop an overall score by subjectively weighing plus and minus deviations from the standard.

Grafius[1] emphasized that this exercise should be based on data only for those traits that are economically important. Data on components of yield should not be included unless they are important in their own right. If selection is to be based on components of a complex trait, as well as the complex trait itself, their relative importance can be assessed by regressing the complex trait on each of the components and multiplying the partial regression coefficients by the economic value of the complex trait. Where a complex trait, such as yield, is a multiplicative function of its components, both the complex trait and its components should be expressed on a logarithmic scale. In this way, the complex trait can be expressed as a linear function of its components.

To illustrate this point more clearly, consider that the standardized partial regression coefficient for the regression of overall excellence score on the logarithm of grain yield is b_v. Then, if the standardized partial regression of the logarithm of grain yield on the logarithm of seed weight, for example, is b_w, the relative economic value of seed weight should be recorded as $b_v b_w$ per standard deviation of the logarithm of seed weight.

Grafius[1] also suggested that traits which do not show consistent performance from one environment to another should not be heavily weighted in the breeding program. He recommended that the standardized partial regression coefficients be multiplied by an estimate of the inter-season correlation of the corresponding trait. In this way, traits with low heritability would receive proportionately less weight in the choice of parents.

B. Proposal by Pederson

Pederson[2] proposed that the relative importance of traits could be developed by specifying an ideal value for each trait, as well as a maximum preferred deviation from that ideal. The objective would be to produce a population whose mean for trait i is $I_i \pm D_i$, where I_i is the ideal value for trait i, and D_i is the maximum preferred deviation. Pederson indicated that this type of scheme is equivalent to giving a weight of $w_i = 1/D_i^2$ to each trait.

This approach is similar to that proposed by Grafius and Adams,[3] and described in some detail by Grafius.[4] These authors recommended that the data for the ith trait be transformed according to

$$X_i' = 1.00 + \frac{X_i - \overline{X}}{10 R_i \overline{X}}$$

where R_i is the acceptable range over which a genotype may vary for the ith trait. For most purposes, the transformation

$$X_i' = \frac{X_i}{R_i \overline{X}} \text{ or } X_i' = \frac{X_i}{R_i}$$

will serve the same purpose as that given by Grafius. Likewise, in the approach described by Pederson,[2] a transformation $X'_i = X_i/D_i$ will give the desired results. The acceptable range of Grafius and Adams[3] serves the same purpose as the maximum preferred deviation of Pederson.[2] In both cases, smaller values indicate greater importance of the corresponding trait.

The regression approach proposed by Grafius[1] will give relative weights similar to those developed by specifying maximum preferred deviations only if $1/D_i^2$ is equal to the partial regression coefficient for the regression of overall worth on trait i. This relationship will be true if overall worth is defined as the sum of X_i/D_i^2 over all traits considered in the overall score. In the regression approach, regression coefficients can be modified to reflect differences in heritability. It should be possible to make similar adjustments in the Pederson[2] approach. Maximum preferred deviations should be increased for traits with lower heritabilities.

With reference to the data presented in Table 1, maximum preferred deviations of 7.1, 3.8, 5.9, 6.2, and 4.2 for traits A, B, C, D, and E, respectively, would give weights equivalent to weights per standard deviation of 0.11, 0.45, 0.57, 0.18, and 0.57 found by the Grafius[1] method.

Grafius[1] recommended that absolute values of the regression coefficients be used as weights since both positive and negative deviations from the ideal values can often be considered to be equally undesirable. This further serves to show the similarity between the two methods. If, in the regression approach, one uses the concept of acceptable range or maximum preferred deviation, both approaches will lead to similar estimates of relative importance.

III. SPECIFYING AN IDEAL GENOTYPE

Grafius[1] suggested that an ideal genotype represented a statement of what is wanted from a given population. As such, the value for each trait will not likely be the maximum attainable for that trait. Rather, an ideal value should reflect a practical level that can be attained by using the parents that are at hand. One would begin the exercise of specifying an ideal genotype by looking at the performance of one or several standard cultivars and at the range of expression within the sample of potential parents. The ideal values should be specified only for those traits which are economically important within the crop.

Pederson[2] and Whitehouse[5] have also discussed the concept of specifying an ideal genotype. An ideal genotype serves as a clear definition of the objectives of the breeding program and therefore plays an important role in choosing suitable parents for that breeding program.

IV. LEAST-SQUARES APPROACH TO CHOOSING PARENTS FOR A BREEDING PROGRAM

The method proposed by Pederson[2] for choosing parents will be discussed first for pairs of parents and then for more complex situations. Consider n potential parents each evaluated for m economically important traits. Let x_{ij} be the mean value for the jth trait measured on the ith potential parent. Develop weights for each trait according to the methods outlined above. It is assumed that ideal values, I_i, have been specified for each of the m traits.

The objective is to choose two parents which, when crossed together in specified proportions, will give a population whose mean values show the smallest weighted sum of squares of deviations from the ideal values. If parents i and i′ are crossed so that parent i contributes a portion p to the genotypic constitution of the progeny population and parent i′ contributes a portion $(1 - p)$, the expected mean for trait j of the progeny population is $C_j = px_{ij} + (1 - p)x_{i'j}$. One must choose a value of p which minimizes the weighted squares of differences between the C_j and the ideal values I_j, i.e.,

$$\sum w_j(C_j - I_j)^2 = \sum w_j[px_{ij} + (1 - p)X_{i'j} - I_j]^2$$

The minimum value for this expression will occur when

$$p = \frac{\sum w_j(X_{ij} - X_{i'j})(I_j - X_{i'j})}{\sum w_j(X_{ij} - X_{i'j})^2}$$

The degree of relationship between the mean of a population formed by crossing parent i with parent i′ in the proportions of p:(1 − p) is measured by the deviation sum of squares, i.e., by

$$\sum w_j[pX_{ij} + (1 - p)X_{i'j} - I_j]^2$$

The procedure for selecting the best pair of parents would be to calculate p for each possible pairwise combination of parents and to choose that pair which has the lowest deviation sum of squares.

In actual practice, one would wish to consider only those values of p and $(1 - p)$ that are realistic. For a single cross, p = 0.5. With backcrossing, proportions of 0.25, 0.125, 0.0625, etc., as well as their complementary values 0.75, 0.875, 0.9375, etc., can be obtained. If a particular proportion suggests the need for two backcrosses to one or the other of the parents, one should recalculate the deviation sum of squares to see if a simpler crossing scheme could be expected to give essentially the same results.

Pederson[2] has described the method for complex crosses involving up to k parents simultaneously. In this case, the proportions p_i, i = 1 to k − 1, of each of the parents are determined by minimizing the expression

$$\sum w_j[\sum p_i X_{ij} - I_j]^2$$

subject to the restriction that $\Sigma p_i = 1.0$. The following set of simultaneous equations would have to be solved to give the required proportions of each of the potential parents in a cross.

$$\sum p_h \sum w_j(X_{ij} - X_{kj})(X_{hj} - X_{kj}) = \sum w_j(X_{ij} - X_{kj})(I_j - X_{kj})$$

The value for p_k is given by $p_k = 1 - \Sigma p_i$ where the summation is over values of h from 1 to k − 1. In each case, one would calculate the residual sum of squares and choose that set of parents having the lowest residual sum of squares. This method can be illustrated with reference to the simulated data in Table 1.

For the five traits listed in Table 1, assume that the ideal values are determined to be 32.0, 47.0, 430, 47.0, and 99.0, respectively, and that the corresponding preferred maximum deviations are given as 7.1, 3.8, 5.9, 6.2, and 4.2. To evaluate the potential of parents 1 and 2 (the first two in Table 1) for meeting these objectives, proceed as follows:

1.   Calculate the weighted sum of squares of differences between the two cultivars.

$$A = (31.7 - 26.4)^2/7.1^2 + (28.1 - 44.0)^2/3.8^2$$
$$+ (401 - 453)^2/5.9^2 + (43.2 - 48)^2/3.5^2$$
$$+ (104.8 - 104.8)^2/5.0^2 = 96.34$$

2.   Calculate the weighted sum of products of the differences between the specified ideal and parent 2 and the differences between parents 1 and 2.

$$B = (32.0 - 26.4)(31.7 - 26.4)/7.1^2$$
$$+ (47.0 - 44.0)(28.1 - 44.0)/3.8^2$$
$$+ (430 - 453)(401 - 453)/5.9^2$$
$$+ (47.0 - 48.0)(43.2 - 48.0)/6.2^2$$
$$+ (99.0 - 104.8)(104.8 - 104.8)/4.2^2 = 31.77$$

3.   Calculate $p = B/A = 31.77/96.34 = 0.330$, and $(1 - p) = 1 - 0.330 = 0.670$
4.   Calculate the residual sum of squares (RSS).

$$RSS = [(p)(31.7) + (1 - p)(26.4) - 32.0]^2/7.1^2$$
$$+ [(p)(28.1) + (1 - p)(44.0) - 47.0]^2/3.8^2$$
$$+ [(p)(401) + (1 - p)(453) - 430]^2/5.9^2$$
$$+ [(p)(43.2) + (1 - p)(48.0) - 47.0]^2/6.2^2$$
$$+ [(p)(104.8) + (1 - p)(104.8) - 99.0]^2/4.2^2 = 7.90$$

If this were the lowest deviation sum of squares, one would conclude that a cross between parents 1 and 2, with parent 1 contributing 33% and parent 2 contributing 67% to the progeny population, would be best. In fact, for the ideal values specified above, evaluation of all 190 possible pairs of parents listed in Table 1 revealed that a cross between parents 5 and 8 would have the smallest deviation sum of squares (0.36), and that the optimum contributions of parents 5 and 8 would be in the ratio 0.35:0.65. With equal contributions of the two parents, the residual sum of squares would be 0.63, close to the minimum value for any other cross. This would give results very nearly the same as for the optimum proportions. One would, therefore, be advised to use a single cross between parents 5 and 8 in order to develop a population whose means are as close as possible to the ideal values specified for this example.

In carrying out the calculations shown above, one will sometimes get estimates of proportions which lie outside the acceptable range of 0 to 1. In these cases, negative estimates are understood to represent estimates of zero contribution for that particular parent.

To evaluate the possibility of using more complex crosses than crosses involving only two parents, similar, but more complex, calculations are required. In effect, if there are to be n parents in a cross, one will calculate $(n - 1)$ sums of the A type given above, as well as $(n - 1)$ sums of type B. These are then combined into $(n - 1)$ simultaneous equations which must be evaluated to give estimates of the first $(n - 1)$ proportionate contributions. Readers are referred to the paper by Pederson[2] for further elaboration on this technique.

The method which Grafius[1] proposed for choosing parents gives quite similar results if the ratio of the standardized partial regression coefficient to standard deviation is equal to the reciprocal of the square of the maximum preferred deviation (i.e., if $b_i/s_i = 1/D_i^2$). For example, consider parents 5 and 8 from Table 1. Grafius recommended that the data be transformed according to

$$X_i' = 1.0 + \frac{b_i}{s_i} (X_i - \overline{X})$$

which gives values of 0.966, 1.429, 1.884, 0.815, and 1.211 for parent 5, values of 1.012, 1.249, 1.542, 1.021, and 0.635 for parent 8, and values of 1.072, 0.952, 1.313, 0.835, and 0.920 for the ideal. The correlations of the transformed values for parents 5 and 8 and the ideal were $r_{58} = 0.68$, $r_{51} = 0.98$, and $r_{81} = 0.92$. These values resulted in the following equations.

$$b_5 + 0.68\, b_8 = 0.98$$

$$0.68\, b_5 + b_8 = 0.92$$

Solving these equations gave estimates of relative optimal contributions of parents 5 and 8 of $b_5 = 0.509$ and $b_8 = 0.575$. These values were then standardized so that they summed to unity. Dividing each by the sum of the two gave optimal contributions of parents 5 and 8 of 0.47 and 0.53, respectively. The coefficient of determination (99.1%) was the highest of all 190 possible two-way crosses. Thus, Grafius' method would give the same overall conclusion as would that of Pederson[2] in this example. The relative contributions of the two parents required to maximize the expected correlation between the cross mean and the ideal genotype (or minimize the deviations) was slightly different in the two methods.

When the weightings are the same in the two methods, the methods of Pederson[2] and Grafius[1] differ in that the Pederson method is based on a model of the expected ideal which considers that the ideal will be zero if all traits are zero. In contrast, the Grafius method has provision for a nonzero intercept. Thus, for Pederson's method, expected $I_j = px_{1j} + (1 - p)x_{2j}$, while for Grafius' method, expected $I_j = a + b_1x_{1j} + b_2x_{2j}$. In both cases, the coefficients p or a, b_1, and b_2 are determined by a weighted least-squares method in which the quantity $\Sigma w_j(I_j - \text{expected } I_j)^2$ is minimized. The Grafius[1] method requires estimation of $n + 1$ parameters while the Pederson[2] method requires the estimation of $n - 1$ parameters where n is the number of parents to be included in a cross. The number of parameters in Pederson's approach is reduced by two by incorporating the restriction that the contributions of parents must sum to unity into the equation, and by setting the intercept at zero.

V. CHOOSING PARENTS TO GIVE MAXIMUM GENOTYPIC VARIABILITY

Bhatt[6] indicated that genotypic variation for grain yield in wheat, *Triticum aestivum* L., was greater in crosses for which the parents showed greater divergence as measured by a multivariate generalized distance statistic. The multivariate distance between parents had been measured on the basis of data collected on days to ear emergence, plant height, tiller number, kernel weight, grain yield per plant, and grain yield per unit area. Bhatt[6,7] recommended the multivariate distance analysis as a method for choosing parents to maximize genotypic variability within subsequent crosses. This method seems to warrant serious consideration in planning long-term selection programs.

Bhatt[7] used the inverse of the error covariance matrix from a randomized complete block analysis of covariance as the basis for weighting the traits in the calculation of multivariate distance between each pair of parents. Thus, the multivariate generalized distance between two genotypes, i and i', based on data for traits j, j = 1 to m, can be defined as

$$D^2_{ii'} = \sum_j \sum_{j'} E^{jj'} (X_{ij} - X_{i'j})(X_{ij'} - X_{i'j'})$$

where summations are over j and j' from 1 to m, and $E^{jj'}$ is the jj'th element of the inverse of the error covariance matrix. For example, consider that the error covariance matrix for the data in Table 1 is

$$\mathbf{E} = \begin{bmatrix} 2.71 & -1.90 & -5.66 & 0.51 & -0.26 \\ -1.90 & 6.01 & 6.60 & 0.18 & -1.65 \\ -5.66 & 6.60 & 76.30 & 1.27 & 1.68 \\ 0.51 & 0.18 & 1.27 & 7.09 & -1.77 \\ -0.26 & -1.65 & 1.68 & -1.77 & 5.16 \end{bmatrix}$$

with inverse

$$\mathbf{E}^{-1} = \begin{bmatrix} 0.549 & 0.162 & 0.026 & -0.033 & 0.060 \\ 0.162 & 0.258 & -0.013 & 0.008 & 0.098 \\ 0.026 & -0.013 & 0.016 & -0.007 & -0.011 \\ -0.033 & 0.008 & -0.007 & 0.159 & 0.058 \\ 0.060 & 0.098 & -0.011 & 0.058 & 0.251 \end{bmatrix}$$

The generalized distance between parents 1 and 2 in Table 1 can be calculated as

$$\begin{aligned} D^2_{12} = & (0.549)(31.7 - 26.4)^2 + (0.258)(28.1 - 44.0)^2 \\ & + \ldots + (0.251)(104.8 - 104.8)^2 \\ & + (2)(0.162)(31.7 - 26.4)(28.1 - 44.0) + \ldots \\ & + (2)(0.058)(43.2 - 48.0)(104.8 - 104.8) = 63.7 \end{aligned}$$

Such a measure of distance will be very sensitive to differences in variability of the traits included in the calculation. The data could be standardized to reduce this problem. However, it would also be valuable if traits of greater relative importance could have relatively more influence on the distance measurement. For this reason, it is suggested that values for each trait be multiplied by weights developed by one or the other of the methods described in Section II, prior to calculating the generalized distance coefficients. This new measure of distance would be

$$D^2_{ii'} = \sum_j \sum_{j'} w_j w_{j'} E^{jj'} (X_{ij} - X_{i'j})(X_{ij'} - X_{i'j'})$$

For example, if one were to use the Grafius[1] approach, weights for traits A, B, C, D, and E (Table 1) would be $0.11/5.5 = 0.0200$, $0.45/6.5 = 0.0692$, $0.57/20.0 = 0.0285$, $0.18/7.0 = 0.0257$, and $0.57/10.0 = 0.0570$, respectively. The distance between parents 1 and 2 would be estimated by

$$D_{12}^2 = (0.0200)\ (0.549)(31.7 - 26.4)^2 + \ldots$$

$$+ (2)(0.0200)(0.0692)(0.058)(31.7 - 26.4)(28.1 - 44.0) + \ldots$$

$$= 0.268$$

A similar approach could be used for relative importance specified in terms of Pederson's maximum preferred deviation,[2] D_i. In this case, one would use weights $w_i = 1/D_i^2$.

The weighted distances between each of the 190 possible pairs of parents listed in Table 1 were calculated. Generalized distance coefficients varied from 0.009 for parents 8 and 14 to 0.581 for parents 1 and 16. This suggests that, for this example, a cross between parents 1 and 16 would result in maximum genotypic variation for overall genotypic worth.

As outlined, this procedure is useful only if one is considering making crosses between two parents. For long-term selection programs, it would be a good idea to use more than two parents to establish the breeding program. In this case, a clustering technique could be used to group potential parents based on the weighted distances between them. One could then choose one parent from each of three or more divergent clusters. Bhatt[7] conducted such a clustering analysis. Several computer packages contain programs for this purpose.

VI. SUMMARY

This chapter has been devoted to the description and illustration of techniques which should be valuable in choosing parents for breeding programs. Relative importance can be determined in either of two subjective ways. In one method, one must specify a subjective score for overall genotypic worth of each of a sample of potential parental genotypes. Multiple regression of the overall score on each of the economically important traits will then give an estimate of the relative importance of each trait. The alternative method is to subjectively evaluate each trait and to specify an acceptable range or maximum preferred deviation for that trait. The narrower the acceptable range, the more important the trait.

An ideal genotype is a succinct way of specifying the objectives of a breeding program. Specifying an ideal is a subjective exercise. However, once an ideal has been specified and estimates of relative importance have been determined for the important traits in the crop, objective methods are available for choosing the best set of parents for a breeding program.

In short-term breeding programs, it is important to choose parents so that the means of their progeny will be as close as possible to the specified ideal genotype. The vector analysis of Grafius,[1] or the least-squares method of Pederson,[2] will give similar results in most instances. The least-squares method has been described in some detail. If several pairs of parents are expected to produce crosses which are equally close to the ideal, the plant breeder would be wise to choose that pair or set which would be expected to produce the greatest genotypic variation in the progeny population.

For long-term selection programs, it is important to start with a population which has sufficient genotypic variability to sustain response to selection over the course of the breeding program. For this purpose, a weighted version of the generalized distance statistic appears to be useful for identifying pairs of parents which are genotypically diverse. For more complex crosses, some form of clustering may be required in order to group potential parents into diverse groupings.

REFERENCES

1. **Grafius, J. E.,** A Geometry of Plant Breeding, Michigan State University Agricultural Experiment Station Res. Bull. No. 7, East Lansing, 1965.
2. **Pederson, D. G.,** A least-squares method for choosing the best relative proportions when intercrossing cultivars, *Euphytica,* 30, 153, 1981.
3. **Grafius, J. E. and Adams, M. W.,** Eugenics in crops, *Agron. J.,* 52, 519, 1960.
4. **Grafius, J. E.,** Vector analysis applied to crop eugenics and genotype-environment interaction, in Statistical Genetics and Plant Breeding, No. 982, Hanson, W. D. and Robinson, H. F., Eds., National Academy of Science, National Research Council, Washington, D.C., 1963, 197.
5. **Whitehouse, R. N. H.,** An application of canonical analysis to plant breeding, Proc. Fifth Cong. EU-CARPIA, Milano, 1968, 61.
6. **Bhatt, G. M.,** Comparison of various methods of selecting parents for hybridization in common bread wheat *(Triticum aestivum* L.), *Aust. J. Agric. Res.,* 24, 457, 1973.
7. **Bhatt, G. M.,** Multivariate analysis approach to selection of parents for hybridization aiming at yield improvement in self-pollinated crops, *Aust. J. Agric. Res.,* 21, 1, 1970.

Chapter 10

COMPUTER PROGRAMS FOR THE STUDY AND APPLICATION OF SELECTION INDICES

I. INTRODUCTION

A computer can be a useful tool for developing a better understanding of complex genetic and statistical problems. The ability to simulate sample data for a specified genetic model, or to simulate data from a specified multivariate normal distribution, and the ability to subsequently analyze the simulated data, can often extend one's understanding of a complex problem. Being able to specify the underlying nature of a population, and to then observe the results of analysis of a sample of data from that population, is a type of experimentation that will often be more convincing than any theoretical argument.

Use of computers is almost essential for analyzing the type of multivariate data that is used with selection indices. Few would be willing to spend hours using a hand calculator to calculate mean squares and cross-products from a replicated test of 50 or so genotypes.

Current versions of popular statistical software packages often do not provide the estimates required for selection indices. Moreover, some general programs are severely limited in the size of problem that can be analyzed, so much so that they are of questionable value to plant breeders working with selection indices.

A recent phenomenon in agricultural research is the availability of microcomputers and their use in analyzing plant breeding data. Software that is particularly adapted to plant breeding problems, especially those for selection indices, is not yet available for use on microcomputers. Because of the magnitude of the computing task, and because programs which would prove to be useful to plant breeders are not generally available, this chapter attempts to provide some information on how a plant breeder might carry out the analyses required for using selection indices.

II. SIMULATION OF GENETIC AND MULTIVARIATE DATA

This author is not aware of any computer programs which are designed specifically for simulating genetic data, although several books have been written on the subject.[1,2] An important component of any genetic simulation program is a random number generator. Most computer systems will include a program which will generate a long series of random numbers distributed uniformly over the interval from 0 to 1. These series are best called "pseudo-random" since each new number in the sequence is uniquely determined by the preceding number. In fact, two identical sets of pseudo-random numbers can be generated by using the same seed to start each series.

Some random number generators may not be very useful for simulation. There may be a serial correlation between consecutive numbers which can result in an undesirable pattern in the simulation results. Or the distribution of random numbers may not be truly uniform. These effects can have serious consequences on simulation results. For serious research based on computer simulation, a great deal of effort needs to be devoted to the choice and evaluation of a random number generator.

The randomness of a random number generator is not so serious if the simulated data is to be used for illustrative purposes rather than for research purposes. While it is still desirable to have a generator which gives a uniform distribution, a few quick tests of the generator will usually be sufficient to determine if it gives data that are satisfactory for illustrative purposes. A short program, such as that listed in Appendix 1, can be used to evaluate a

Table 1

**MEANS, STANDARD DEVIATIONS, AND
GOODNESS-OF-FIT TESTS FOR 10 SAMPLES OF 500
UNIFORM RANDOM NUMBERS**[a]

Sample number	Mean	Standard deviation	Goodness-of-fit (chi-squared)[b]
1	0.508	0.300	7.6
2	0.503	0.290	8.8
3	0.495	0.283	8.0
4	0.519	0.285	8.6
5	0.503	0.282	4.9
6	0.482	0.288	8.9
7	0.524	0.283	13.8
8	0.501	0.301	6.2
9	0.487	0.287	8.8
10	0.503	0.295	5.8
Average	0.503	0.289	8.1
Expected value	0.500	0.289	

[a]   Numbers generated using the RND function of cartridge BASIC for the IMB PCjr microcomputer.

[b]   Chi-squared test for goodness-of-fit to expected frequencies of 50 in each of ten classes; degrees of freedom = 9; 5% critical value = 16.9.

random number generator. The program in Appendix 1 was used to evaluate the uniform random number generator available in cartridge BASIC on an IBM PCjr microcomputer. In 10 sets of 500 random numbers, means were close to the expected value of 0.5, and standard deviations were close to the expected value of 0.2887 (Table 1). In all 10 sets, the distributions of random numbers gave good fits to the expected uniform distribution.

Simulation of genetic models can be a very complex process, especially when simulation is of two or more quantitative traits. Simulation programs could be designed to allow the user to specify numbers of genes affecting each trait, probabilities of recombination between genes at different loci, sizes of additive, dominant, and epistatic effects, as well as levels of environmental variation and covariation. Experience has shown that such comprehensive programs are difficult to use because of the need to specify so many different parameters. In practice, one must compromise between programs which are capable of simulating realistically complex genetic models, and programs which offer more limited choice but are easier to use.

As an example of genetic simulation, the data used to demonstrate analysis of heterogeneous and homogeneous populations (Chapter 4, Section VIII) were simulated using a program similar to that listed in Appendix 2. This program is suitable for running on a microcomputer or a mainframe computer and is written in BASIC. It is limited to the simulation of genotypic models in which there is no linkage between genes at different loci and in which all alleles are additive in their effects. Each allele may affect any or all traits and may have different effects for each trait. This facilitates simulation of models where one allele has a large effect and others have small effects on a particular trait, or models where an allele can have pleiotropic effects on two or more traits.

This program allows the user to specify gene frequencies at each locus as well as the average level of inbreeding of the population that is being simulated. Alleles at the ith locus may be designated as A_i and a_i and the frequencies of the three possible genotypes at that locus, viz. A_iA_i, A_ia_i, and a_ia_i, are set equal to $p_i^2 + Fp_i(1 - p_i)$, $2p_i(1 - p_i)(1 - F)$, and $(1 - p_i)^2 + Fp_i(1 - p_i)$, respectively, where p_i is the frequency of the allele A_i in the population and F is the inbreeding coefficient.

For the jth trait, the genotypic effects for genotypes at the ith locus are g_{ij} for A_iA_i, 0 for A_ia_i, and $-g_{ij}$ for a_ia_i. The genotypic value for any trait is given by the sum of the genotypic effects at all loci which carry genes affecting that trait, plus the overall mean value for that trait. If $F = 1$, the population is considered to be completely inbred and all genotypic values are set equal to the mean values specified for each trait. A population consisting of F_1 plants from a cross between two homozygous parents can be simulated by setting $F = 1$ and specifying the average genotypic values to be equal to the corresponding mid-parent values.

The program, as listed in Appendix 2, is designed to simulate phenotypic values for several plants in each of several plots. For this reason, one must indicate the environmental variances and covariances which apply to plot-to-plot variation, as well as those which apply to plant-to-plant variation within plots.

Since the primary purpose of this program is to simulate data for heterogeneous and homogeneous populations, simulated data are written into a file in a manner suitable for subsequent analysis by other programs. The simulated data considered in Chapter 4, Section VIII consisted of phenotypic values for two homozygous parents, and their F_1 and F_2 progeny. Ten plots of each parent, 20 of the F_1, and 40 of the F_2, each consisting of 5 plants, were simulated. Two traits, controlled by a total of 15 loci, were considered. This program can be used for considering similar models or can be modified for simulation of unrelated problems. It serves primarily as an example of elementary genetic simulation. For those who are serious about genetic simulation as a research tool, they would be advised to read books, such as that by Fraser and Burnell,[1] which give detailed descriptions of the methods used for genetic simulation, as well as books, such as that by Knuth,[3] dealing with the evaluation and choice of random number generators.

In some of the applications discussed in previous chapters, it was necessary to have sample estimates from populations in which the genotypic and phenotypic values were considered to be normally distributed. A short program (Appendix 3) was developed for this purpose. The program is written in BASIC and combines several complex transformations in order to convert independent, uniformly distributed random numbers into independent normal deviates, and then into correlated normal deviates (See Naylor et al.[4]). The user must indicate the number of traits that are to be simulated, as well as the size of the sample that is to be generated. Then, for each trait, the user must specify a mean value, the genotypic variance, and the heritability. The user must also specify the genotypic and environmental correlations for each pair of traits.

The program will generate the specified sample and write the estimated genotypic and phenotypic variances and covariances on a diskette file for subsequent use. This program is particularly useful for investigating various characteristics of correlated normal variables. The subroutines included in the program should prove useful for other applications.

III. ESTIMATING VARIANCES AND COVARIANCES FROM BALANCED DESIGNS

Many commercial statistical software packages include programs for analyzing multivariate data. However, not all of them are capable of providing estimates of mean squares and mean cross-products as required for developing selection indices. For some of these programs, mean squares and mean cross-products are intermediate calculations in the overall analysis and no facility is provided for their display. In other cases, programs for multivariate analysis have been generalized to handle both balanced and unbalanced data sets. In these cases, the size of the experiment that can be analyzed is sometimes too limited to be of use in developing selection indices.

Several popular statistical packages were tested for their ability to provide estimates of mean squares and mean cross-products in a balanced design. The test data set was the

Table 2
SAS CONTROL STATEMENTS REQUIRED TO ANALYZE A REPLICATED
N. C. DESIGN II EXPERIMENT[a]

```
DATA;
  INFILE DATAIN;
  INPUT P 1-3 S 4-6 R 7-9 M 10-12 F 13-15 Y1 19-25 Y2 26-32
PROC ANOVA;
  CLASSES P S R M F;
  MODEL Y1-Y2 = P R(P) S P*S R*S(P) M(S) F(S) M*F(S) P*M(S) P*F(S) P*M*F(S);
  MANOVA H = M(S) /PRINTH;
  MANOVA H = F(S) /PRINTH;
  MANOVA H = M*F(S) /PRINTH;
  MANOVA H = P*M(S) /PRINTH;
  MANOVA H = P*F(S) /PRINTH;
  MANOVA H = P*M*F(S) /PRINTH PRINTE;
```

[a]   See text for description of data set.

simulated data for the North Carolina Design II experiment described in Chapter 4, Section VI. The simulated experiment consisted of 20 sets of 24 full-sib families (four males each crossed with six females). Each full-sib family was replicated three times in each of two environments. Data for two traits were simulated for each of the 2880 experimental units. This data set is quite large for a Design II experiment, but the total number of observations (5760) is only moderately large when one is dealing with multiple trait experiments.

The data were first stored in a file consisting of 2880 records each with seven fields. Contents of the seven fields were environment number, set number, replication (within environment) number, male (within set) number, female (within set) number, value for trait A, and value for trait B. This data file was then analyzed by using the SAS (Statistical Analysis System[5]) and the SPSS[X] (Scientific Programs for the Social Sciences[6]) statistical programs.

For SAS,[5] the ANOVA procedure was used to analyze the data according to the model given in Chapter 4. To print the mean squares and the mean cross-products for those sources of variation for which estimates are required, the MANOVA command was used in combination with the PRINTH switch and the PRINTE switch (Table 2). The SAS program actually prints the sums of squares and cross-products and these must be divided by the corresponding degrees of freedom to obtain the required mean squares and mean cross-products. The output from SAS includes a series of tests on the cross-products matrix. Most of the output is unnecessary in estimating variances and covariances for use in selection indices.

With the SPSS[X] statistical package,[6] the MANOVA command can be used to calculate and print sums of squares and cross-products associated with each term in the model. The PRINT subcommand (See Table 3) can be used to print the sums of squares and cross-products associated with each term in the model as well as the mean squares and mean cross-products for the error term.

The MANOVA program of SPSS[X] is designed for multivariate analysis of unbalanced data. As a result, its use for estimating mean squares and mean cross-products for balanced designs is severely limited. For example, with sets of 24 full-sib families evaluated in three replications in each of two environments, only two sets could be analyzed on a Digital Equipment Systems 2060 computer running SPSS[X] Release 1.0 for DEC-20. For this particular design, there are a total of 104 parameters that must be estimated if two sets are analyzed, 276 if three sets are analyzed, and 941 if all 20 sets are analyzed. If the design is not balanced, the design matrix must include 10,816 elements for two sets, 76,176 elements

Table 3
SPSS^X CONTROL STATEMENTS REQUIRED TO ANALYZE A REPLICATED
N. C. DESIGN II EXPERIMENT[a]

TITLE SUMS OF SQUARES AND CROSS-PRODUCTS
DATA LIST /P 1-3 S 4-6 R 7-9 M 10-12 F 13-15 Y1 16-25 Y2 26-32
MANOVA Y1,Y2 BY P(1,2) S(1,20) R(1,3) M(1,4) F(1,6)/
   PRINT = SIGNIF(HYPOTH)/
   PRINT = ERROR(COV)/
   DESIGN = P, S, R W P BY S, M W S, F W S, M BY F W S, P BY M W S, P BY F W S, P BY M BY F
   W S/
BEGIN DATA

[a]   See text for description of data set.

for three sets, and 885,481 elements for all 20 sets. With a balanced design, the design matrix consists only of the diagonal elements of the matrix and can be stored in 104, 276, or 941 elements, respectively. It is for this reason that the test data set could be analyzed using the balanced ANOVA procedure in SAS, but could not be analyzed by using the unbalanced MANOVA command of SPSS^X.

The following technique may be used to estimate mean squares and mean cross-products by using any program which is capable of analyzing a balanced factorial design. If A and B are two variables, analysis of variance of each of them, as well as of the new variable C = A + B can be used to estimate mean squares and mean cross-products. For any source of variation, the mean squares are given by the analyses of A and B. The mean cross-product can be estimated by [mean square (A + B) − mean square (A) − mean square (B)]/2. This approach was used to develop the FORTRAN program listed in Appendix 4. The program is based on the IMSL (International Mathematical and Statistical Library[7]) subroutine AGBACP for balanced analysis of variance, and can only be used in conjunction with that subroutine. This program was also tested with the test data set and gave results identical to those obtained by using SAS.[5]

With the development of microcomputers, researchers may wish to carry out calculations for selection indices on these machines. Generally, the programming language will be a version of BASIC, and internal random access memory will be limited to 64,000 bytes. A minimum requirement is that such machines have at least one diskette drive so that data can be read from or written to a diskette or other storage medium. In these applications, computing time is not likely to be an important consideration. These microcomputers are capable of analyzing data sets of sufficiently large size to be of use to plant breeders in the development and application of selection indices.

The program listed in Appendix 5 was developed for estimating mean squares and mean cross-products in balanced designs. The program is based on a method described by Hemmerle,[8] but differs from it in several important aspects. The new program calculates effects corresponding to only one item in the model at one time, and combines them into estimates of the corresponding mean squares and mean cross-products. A second pass, including reading the data from the diskette file, is required for estimating the effects and mean squares of the second term, and so on. In this way, one may estimate mean squares and cross-products only for those items that are required for subsequent estimation of appropriate genotypic and phenotypic variances and covariances. This approach allows fairly large data sets to be analyzed within a limited amount of random access memory.

The computational methods used in this program can be described by reference to the model for a North Carolina Design II experiment as described in Chapter 4. This experiment consisted of five factors; environment (1 or 2), replications within environment (1, 2, or 3), set (1 to 20), males within sets (1 to 4), and females within sets (1 to 6). The model for the analysis is given by

$$P_{ijklm} = \mu + P_i + S_j + PS_{ij} + R_{ijk} + M_{jl} + F_{jm} + MF_{jlm}$$
$$+ PM_{jil} + PF_{jim} + PMF_{jilm} + E_{ijklm}$$

where the model effects P, S, R, M, and F represent the effects of environments, sets, replications within sets and environments, males within sets, and females within sets, respectively, and E stands for the experimental error. The subscripts i, j, k, l, and m correspond to the five factors and are associated with P, S, R, M, and F, respectively. The term PM_{jil} is written with subscripts jil as a reminder that it is the interaction between the ith environment and the lth male within the jth set of full-sib families. The subscripts i and l are associated directly with the effects P and M while the subscript j is a floating subscript which indicates that the PM interaction effects are nested within the jth set.

For use in a computer program, the model must be modified into a computer readable form. For the programs listed in appendices of this book, the model should be written as

P(I)S(J)PS(IJ)M(JL)F(JM)MF(JLM)PM(JIL)PF(JIM)PMF(JILM)E(IJKLM)

In this method, subscripts are enclosed in brackets. Where a term contains more subscripts than there are effects, it will be assumed that the last subscripts in the subscript list are those which are associated with the corresponding effects. The last term in the model must be the error term E and must contain a complete list of subscripts.

To use the program in Appendix 5, one would specify the name of the file that contains the data. On prompting from the program, the user would specify that there are five factors and two variables, and that there are two levels for the first factor, 20 for the second, three for the third, four for the fourth, and six for the fifth. Then, to estimate the mean squares and cross-products for, say, the "environment by females within sets" source of variation (PF_{jim}), the user would indicate that the first subscript, i, is associated with the term, that the second subscript, j, is a floating subscript, that the third subscript, k, is not involved in this term, that the fourth subscript, l, is not involved, and that the fifth subscript, m, is associated with the term.

The program will read the data and calculate means for the two variables over the two subscripts which are not included in this particular term, i.e., calculate $Y_{ij..m}$ for each variable. Then, for each of the subscripts in the term, provided they are associated with an effect in the term, means are calculated for that subscript and subtracted from the means or deviations of the previous step. Thus $Y_{ij..m}$ will be replaced by $(Y_{ij..m} - Y_{i....})$, which will in turn be replaced by $[(Y_{ij..m} - Y_{i....}) - (Y_{....m} - Y_{....})]$. These are the effects corresponding to the PF_{jim} term. The effects for each variable are either squared or multiplied, summed, and multiplied by the products of the levels of the subscripts not included in this term, to give the final estimates of the mean squares and mean cross-products for this term.

This program requires a matrix in random access memory with number of rows equal to the number of data points (2880 in the test case) plus the maximum of the number of levels of the five factors (20 in the test case) and number of columns equal to the number of variables (2 in the test case). Allowing for the storage requirements of the program, and recognizing that each single precision number requires 4 bytes, there is enough room in 64,000 bytes of random access memory for the work matrix to hold about 12,000 numbers. The test case required a matrix of 2900 rows by 2 columns for a total storage requirement of 5800 numbers. It should be possible to handle up to four variables with this large a data set. More variables could be handled with a smaller data set.

The program is quite slow in comparison to large computers. Computing times of approximately 10 minutes were required for each term in the analysis of the test data set on an IBM PCjr computer. Since a total of seven terms are required for the estimation of genotypic and phenotypic variances and covariances, total computing time for the test data set would be something over one hour on this particular microcomputer.

Appendix 6 contains the listing for a second program for calculating mean squares and mean cross-products for balanced designs. This program is based on an algorithm developed by Howell[9] in which a series of normalized orthogonal transformations are used to transform the data vector into a vector which includes a correction factor, all main effects, and all interaction effects. Some of the computational methods described by Hemmerle[8] are then used to gather the appropriate terms and synthesize the required mean squares and cross-products. This program requires somewhat more workspace than the one listed in Appendix 5 (approximately one extra column in the work matrix). However, the program has the advantage of providing the complete analysis of covariance and does not require that the user be present to initiate calculations of each new term. For the test data set, this program required approximately 2 hours and 20 minutes to calculate the complete analysis of covariance on an IBM PCjr microcomputer.

Either of these programs could be modified to run on other microcomputers or on mainframe computers. The BASIC used in developing these programs consists of a standard subset of BASIC statements. Statements for opening and closing sequential files, statements for clearing the screen or turning the key definitions off, and perhaps statements dealing with some of the string handling functions would likely require modification. It may also be necessary in some instance to shorten variable and array names to two characters. It would not be difficult for anyone having programming knowledge of both BASIC and FORTRAN to convert these programs to FORTRAN.

In analyzing experiments for developing selection indices, most effects in the analysis model are considered to be random effects and it is not difficult to write down the expectations for the mean squares of a balanced design. If using the programs described above for other purposes, developing expectations of mean squares and deciding on appropriate F-tests may not be so straightforward. To assist in these cases, the program listed in Appendix 7 was developed from rules given by Hicks.[10] The user specifies the model as outlined above, and indicates the levels of each factor and whether the effects of each factor are to be considered as a random sample from some reference population, or as a fixed sample about which inferences are to be made. The program will then calculate and print the expectations of mean squares for each term in the model.

Once mean squares and mean cross-products have been calculated, relatively simple calculations are required to combine them into estimates of genotypic and phenotypic variances and covariances. A program such as that listed in Appendix 8 can be used to calculate the required linear functions. In this program, a few statements have been added to calculate the approximate standard errors (see Chapter 4) of the estimates of genotypic and phenotypic variances and covariances, as well as estimates of the heritability of each trait and its standard error. Calculation of the appropriate standard errors provides some information about the reliability of the estimated variances and covariances.

The BASIC program listed in Appendix 9 can be used for further evaluation of the reliability of estimated genotypic and phenotypic variance-covariance matrices. The program carries out a transformation of the two matrices as outlined by Hayes and Hill.[11] The program is based on algorithms developed by Martin and Wilkinson,[12] Martin et al.[13] and Bowdler et al.[14] The discussion in Chapter 4, Section IX provides a guideline for interpreting the output from this program.

IV. CALCULATING SELECTION INDICES

Any computer program which will allow the user to invert and multiply matrices will suffice for estimating index coefficients from estimated genotypic and phenotypic variances and covariances. Matrix handling programs may not be readily available on microcomputers. In this case, the program listed in Appendix 10 can be used to calculate index coefficients for an estimated optimum index. The program will read the genotypic and phenotypic covariance matrices from a file, prompt the user to specify the relative economic weights for the various traits, and calculate and print the index coefficients. The program will continue and print estimates of expected response to selection based on the estimated covariance matrices. This option may be useful in exploring the limitations to response in the population from which the estimates have been derived. The matrix inversion routine is based on that described by Miller.[15]

Several methods of developing appropriate index coefficients have been discussed in earlier chapters. The estimated optimum index may not be practical in many cases. Whether index coefficients are developed by this method or estimated by the relative economic worth or the product of economic worth and estimated heritability, the index coefficients must be used as weights to develop the overall index scores for final selection. Any computer program which will allow one to operate on columns of numbers, and to sort several columns based on values in one of the columns, can be used for this purpose. A short BASIC program for this purpose is listed in Appendix 11.

Elston's method[16] of ranking genotypes for several traits can also be carried out with computer programs which facilitate column operations. The listing in Appendix 12 provides a short program for conducting such a weight-free ranking on a microcomputer. Use of this program presupposes that all traits have approximately unimodal, symmetric distributions.

In evaluating parental material, the method described by Pederson[17] seems to be most useful. It is unlikely that existing statistical packages can be modified to carry out this kind of calculation. The program listed in Appendix 13 will read the data for the potential parents from a file and prompt the user to specify an ideal genotype and a preferred maximum deviation for each trait. All possible pairs of potential parents will then be compared to the ideal by Pederson's methods. The program evaluates only those crosses which can be developed in one or two cycles of crossing. These include single crosses and single back-crosses between two parents, three-way crosses, and four-way crosses. With n parents, there are $^nC_2$ (combinations of n items taken two at a time = $n!/\{2![n-2]!\}$) single crosses, 2^nC_2 single backcrosses, 3^nC_3 three-way crosses, and $^nC_4$ four-way crosses. Thus, the program will evaluate 65, 705, 3045, and 8835 different crosses for 5, 10, 15, and 20 potential parents, respectively, and indicate which cross in each of the four classes has a mid-parent value which most closely approaches the ideal genotype.

V. SUMMARY

Application of selection indices to plant breeding requires analysis of multivariate data. Few computer packages provide all of the tools required for these calculations. This chapter reviews some of the more popular statistical packages and shows how they can be used for some of the calculations required in index selection. In addition, programs written in BASIC and designed specifically for application to selection indices are introduced. These programs include programs for simulating genetic and multivariate normal data, for calculating mean squares and cross-products in balanced designs, for calculating genotypic and phenotypic covariances and their standard errors, for evaluating estimated covariance matrices, for calculating selection indices, for ranking individuals according to index scores, and for evaluating potential parents.

REFERENCES

1. **Fraser, A. S. and Burnell, D.,** *Computer Models in Genetics,* McGraw-Hill, New York, 1970.
2. **Crosby, J. L.,** *Computer Simulation in Genetics,* John Wiley & Sons, New York, 1970.
3. **Knuth, D. E.,** *The Art of Computer Programming. Vol. 2. Seminumerical Algorithms,* Addison-Wesley, Reading, Mass., 1969.
4. **Naylor, T. H., Balintfy, J. L., Burdick, D. S., and Chu, K.,** *Computer Simulation Techniques,* John Wiley & Sons, New York, 1966, 97.
5. **Helwig, J. T. and Council, K. A.,** *SAS User's Guide, 1979 Edition,* SAS Institute Inc., Cary, N.C., 1979.
6. SPSS Inc., *SPSS^x User's Guide,* McGraw-Hill, New York, 1983.
7. IMSL Inc., *IMSL Library. Problem-Solving Software System for Mathematical and Statistical FORTRAN Programming. User's Manual,* 9th ed., IMSL Inc., Houston, Tex., 1980.
8. **Hemmerle, W. J.,** *Statistical Computations on a Digital Computer,* Blaisdell, Waltham, Mass., 1967.
9. **Howell, J. R.,** Algorithm 359. Factorial analysis of variance, *Comm. of the ACM.,* 12, 631, 1969.
10. **Hicks, C. R.,** *Fundamental Concepts in the Design of Experiments,* Holt, Rinehart & Winston, New York, 1973.
11. **Hayes, J. F. and Hill, W. G.,** A reparameterization of a genetic selection index to locate its sampling properties, *Biometrics,* 36, 237, 1980.
12. **Martin, R. S. and Wilkinson, J. H.,** Reduction of the symmetric eigenproblem $Ax = \lambda Bx$ and related problems to standard form, *Numer. Math.,* 11, 99, 1968.
13. **Martin, R. S., Reinsch, C., and Wilkinson, J. H.,** Householder's tridiagonolization of a symmetric matrix, *Numer. Math.,* 11, 181, 1968.
14. **Bowdler, H., Martin, R. S., Reinsch, C., and Wilkinson, J. H.,** The QR and QL algorithms for symmetric matrices, *Numer. Math.,* 11, 293, 1968.
15. **Miller, A. R.,** *BASIC Programs for Scientists and Engineers,* SYBEX Inc., Berkeley, Calif., 1981.
16. **Elston, R. C.,** A weight-free index for the purpose of ranking and selection with respect to several traits at a time, *Biometrics,* 19, 85, 1963.
17. **Pederson, D. G.,** A least-squares method for choosing the best relative proportions when intercrossing cultivars, *Euphytica,* 30, 153, 1981.

APPENDIX 1:

PROGRAM FOR TESTING RANDOM NUMBER GENERATOR

PROGRAM NAME: RNDTST.BAS

LANGUAGE: BASIC

PURPOSE:
This program performs simple tests on a uniform random number generator.

DESCRIPTION:
This program will generate a set of random numbers that are uniformly distributed over the interval from 0 to 1. The size of the sample is specified by the user. For each set, the program calculates the mean value (expected to be 0.500) and the standard deviation (expected to be 0.2887), as well as a chi-squared statistic for the fit to a truly uniform distribution. The latter test is based on a subdivision of the interval 0 to 1 into 10 equal intervals and a subsequent test of the observed proportion in each interval against the expected proportion of 0.10.

USER INSTRUCTIONS:
The user loads the program, runs it, and responds to the prompt to indicate sample size. Sample sizes must be 50 or greater. If not, the program will reissue the prompt. For the first sample, the user will also be asked to provide an integer number in the range of -32768 to 32767 which serves as a seed to start the random number generator. If the same seed is used each time the program is run, the same sequence of random numbers will be generated. For each sample, the program will print the mean, the standard deviation, and the frequency distribution of the simulated sample.

Computers differ in how they start a series of random numbers and in how the random number subroutine is called. Such differences may require modification to the segment of program that serves to seed the generator (line 220 to 320) and the line (line 480) that sets the variable X equal to the next random number in the series.

PROGRAM LISTING:
```
 10   REM Program for evaluating a uniform (0,1) random
 20   REM number generator. The operator is prompted to
 30   REM indicate the size of sample to be evaluated.
 40   REM The program generates the sample and the
 50   REM following test statistics are calculated:
 60   REM        a) mean (expected to be 0.5)
 70   REM        b) standard deviation (expected to be 0.2887)
 80   REM        c) frequencies in each of 10 classes
 90   REM        d) chi-squared value for fit to U(0,1) distribution
100   REM
110   REM Clear screen and print heading
120   REM Method for clearing screen is machine dependent
130   REM
140   CLS:KEY OFF
150   PRINT "Program to evaluate random number generator"
160   PRINT
170   REM
```

```
180   REM Prompt for sample size
190   REM
200   INPUT "Enter sample size (> = 50) ";N
210   IF N < 50 THEN 200
220   REM
230   REM Prompt for seed to start random number generator
240   REM Method of seeding will depend upon computer
250   REM
260   PRINT "Enter integer value (− 32768 to 32767) as seed"
270   INPUT "for the random number generator ";S%
280   REM
290   REM Seed the generator (method depends on computer)
300   REM
310   RANDOMIZE S%
320   REM
330   REM Dimension class frequency counters
340   REM
350   DIM F(10)
360   REM
370   REM Zero arrays and accumulators
380   REM
390   FOR J = 1 TO 10
400   F(J) = 0
410   NEXT J
420   AV = 0
430   SD = 0
440   REM
450   REM Generate random numbers
460   REM
470   FOR I = 1 TO N
480   X = RND
490   REM Remove the following statement if you do not want
500   REM numbers printed on screen as they are generated
510   PRINT USING " #.###";X;
520   AV = AV + X
530   SD = SD + X * X
540   J = INT(X * 10)+1
550   F(J) = F(J) + 1
560   NEXT I
570   REM
580   REM Calculate and print mean and standard deviation
590   REM
600   SD = SQR((SD − AV * AV / N)/(N − 1))
610   AV = AV/N
620   CLS:PRINT:PRINT
630   PRINT "Sample size = ";N
640   PRINT USING "Mean        = #.####, expected value = 0.5000";AV
650   PRINT USING "Standard deviation = #.####, expected value = 0.2887";SD
660   PRINT
670   REM
680   REM Calculate Chi-squared and print observed and
```

```
690   REM expected frequencies.
700   REM
710   PRINT "Interval        Observed        Expected        Deviation"
720   CS = 0
730   FOR J = 1 TO 10
740   E = N/10
750   I1$ = "0." + RIGHT$(STR$(J − 1),1) + "0"
760   I2$ = "0." + RIGHT$(STR$(J − 1),1) + "9"
770   D = F(J) − E
780   CS = CS + D*D/E
790   PRINT I1$;" to ";I2$;
800   PRINT USING " #####        #####.#        #####.#"; F(J),E,D
810   NEXT J
820   PRINT
830   PRINT "Chi-squared = ";
840   PRINT USING " #####.# ";CS;
850   PRINT " with 9 degrees of freedom"
860   PRINT "Value greater than 16.9 indicates significance"
870   PRINT
880   INPUT "Another sample (y or n) "; Y$
890   IF Y$ <> "y" AND Y$ <> "Y" THEN END
900   CLS
910   INPUT "Enter sample size (> = 50) ";N
920   IF N < 50 THEN 910
930   GOTO 390
940   END
```

PROGRAM FOR GENETIC SIMULATION OF POPULATIONS

PROGRAM NAME: GENSIM.BAS

LANGUAGE: BASIC

PURPOSE:
  This program can be used to simulate genotypic and phenotypic values of heterogeneous and homogeneous populations.

DESCRIPTION:
  This program is capable of simulating genetic models in which there is no linkage between genes at different loci, and in which there is no interaction between alleles at the same locus nor between alleles at different loci (no dominance and no epistasis). The program simulates two alleles at each locus, and there can be any number of loci affecting any number of traits. The program is designed to simulate the genotypic and phenotypic values of each of a number of plants within each of a number of plots.

USER INSTRUCTIONS:
  To run this program, it must first be loaded into memory and have changes made to certain data statements. The number of traits to be simulated, the number of loci which control those traits, the number of plants per row (or plot), the number of plots, and an identification number for the population must be entered into the data statement at line 180. Beginning at line 320, the user must include one data statement for each trait to be simulated. Each statement will include the genotypic values contributed to that trait by the A_iA_i genotype at each locus. The program includes an example which shows the genotypic values for 15 loci and two traits.
  Genotypic frequencies are entered in data statments at lines 410 and 420. The inbreeding coefficient of the population is entered first, then the frequency of the A_i allele at each locus. Environmental variances and covariances among plants within plots are entered in the data statement at line 510, and those among plots are entered in line 600. In both cases, the program requires NT*NT values where NT is the number of traits being simulated. The means for each trait are entered into the data statement at line 670.
  When the program is run, the user will be prompted to enter an integer number to serve as a seed for the random number generator. The user will also specify the name of a file into which the simulated phenotypic values are to be written. Phenotypic values are printed on the screen at the same time they are written into the diskette file.

PROGRAM LISTING:
```
10 REM Program to simulate data from heterogeneous or
20 REM homogeneous populations. Simulates data for
30 REM NT traits all controlled by NL loci. Program
40 REM simulates phenotypic values for NP plants within
50 REM each of NR rows and stores the data in a disk file.
60 REM
70 REM Data is entered through data statements at lines
80 REM 180, 320, 330, 410, 420, 510, 600 and 670.
90 REM
100 REM****************************************************
```

```
110 REM Clear screen (machine dependent)
120 CLS: KEY OFF
130 REM Segment to read input from data statements
140 REM Read NT (# traits), NL (# loci), NP (# plants/row)
150 REM NR (# rows for this population) and ID (a number
160 REM        to identify this population)
170 READ NT,NL,NP,NR,ID
180 DATA 2,15,5,40,2
190 REM
200 REM dimension variables
210 REM
220 DIM A(NT,NL),F(NL,2),P(NL),EB(NT,NT),EW(NT,NT),CB(NT,NT)
230 DIM CW(NT,NT),AV(NT),G(NT),PH(NT),R(NT),B(NT),W(NT)
240 REM
250 REM read genotypic values for each trait
260 REM
270 FOR IT = 1 TO NT
280 FOR IL = 1 TO NL
290 READ A(IT,IL)
300 NEXT IL
310 NEXT IT
320 DATA 1,1,1,1,1,1,1,1,1,1,0,0,0,0,0
330 DATA 0,0,0,0,0,0,0, − 1, − 1, − 1,1,1,1,1,1,1
340 REM
350 REM read inbreeding coefficient and allele frequencies
360 REM
370 READ FF
380 FOR IL = 1 TO NL
390 READ P(IL)
400 NEXT IL
410 DATA 0.0
420 DATA 0.5,0.5,0.5,0.5,0.5,0.5,0.5,0.5,0.5,0.5,0.5,0.5,0.5,0.5,0.5
430 REM
440 REM Read plant-to-plant environmental variances and covariances
450 REM
460 FOR IT = 1 TO NT
470 FOR JT = 1 TO NT
480 READ EW(IT,JT)
490 NEXT JT
500 NEXT IT
510 DATA 5.0, 1.0, 1.0, 5.0
520 REM
530 REM Read plot-to-plot environmental variances and covariances
540 REM
550 FOR IT = 1 TO NT
560 FOR JT = 1 TO NT
570 READ EB(IT,JT)
580 NEXT JT
590 NEXT IT
600 DATA 2.0, 0.0, 0.0, 2.0
610 REM
```

```
620 REM Read means for each trait
630 REM
640 FOR IT = 1 TO NT
650 READ AV(IT)
660 NEXT IT
670 DATA 15.0,15.0
680 REM
690 REM Prompt for seed for random number generator (machine dependent)
700 REM
710 INPUT "Enter integer ( − 32768 to 32767) to start random numbers ";S%
720 RANDOMIZE S%
730 REM
740 REM Prompt for name of file for output data
750 REM
760 INPUT "Enter name of output file ";NF$
770 REM
780 REM Open file for output (machine dependent)
790 REM Open as 'append' so data can be added to
800 REM existing file if desired.
810 REM
820 OPEN NF$ FOR APPEND AS #1
830 REM
840 REM Calculate genotype frequencies at each locus
850 REM
860 FOR IL = 1 TO NL
870 X = 1 − P(IL)
880 F(IL,1) = X*X + FF*X*(1 −X)
890 F(IL,2) = X*(1 −X) − 2 *FF*X*(1 −X) + F(IL,1)
900 NEXT IL
910 REM
920 REM Calculate transformation for multivariate normal simulation
930 REM Method based on transformation given by:
940 REM Naylor, T. H., Balintfy, J. L., Burdick, D. S.
950 REM and Chu, K., Computer simulation techniques,
960 REM John Wiley and Sons, Inc., New York, NY, 1966.
970 REM
980 FOR IT = 1 TO NT
990 CB(IT,1) = EB(IT,1)/SQR(EB(1,1))
1000 CW(IT,1) = EW(IT,1)/SQR(EW(1,1))
1010 NEXT IT
1020 FOR IT = 1 TO NT
1030 TB = EB(IT,IT)
1040 TW = EW(IT,IT)
1050 FOR JT = 1 TO IT−1
1060 TB = TB − CB(IT,JT)*CB(IT,JT)
1070 TW = TW − CW(IT,JT)*CW(IT,JT)
1080 NEXT JT
1090 CB(IT,IT) = SQR(TB)
1100 CW(IT,IT) = SQR(TW)
1110 NEXT IT
1120 IF NT < 3 THEN 1280
```

```
1130 FOR IT = 3 TO NT
1140 FOR JT = 2 TO IT − 1
1150 TB = EB(IT,JT)
1160 TW = EW(IT,JT)
1170 FOR KT = 1 TO JT − 1
1180 TB = TB − CB(IT,KT)*CB(JT,KT)
1190 TW = TW − CW(IT,KT)*CW(JT,KT)
1200 NEXT KT
1210 CB(IT,JT) = TB/CB(JT,JT)
1220 CW(IT,JT) = TW/CW(JT,JT)
1230 NEXT JT
1240 NEXT IT
1250 REM
1260 REM Start simulation loops
1270 REM
1280 FOR IR = 1 TO NR
1290 REM
1300 REM simulate environmental effects for row IR
1310 REM
1320 GOSUB 1880
1330 FOR IT = 1 TO NT
1340 B(IT) = 0
1350 FOR JT = 1 TO NT
1360 B(IT) = B(IT) + CB(IT,JT)*R(JT)
1370 NEXT JT
1380 NEXT IT
1390 REM
1400 REM Loop for each plant
1410 REM
1420 FOR IP = 1 TO NP
1430 REM
1440 REM Simulate environmental effects for each plant
1450 REM
1460 GOSUB 1880
1470 FOR IT = 1 TO NT
1480 W(IT) = 0
1490 FOR JT = 1 TO NT
1500 W(IT) = W(IT) + CW(IT,JT)*R(JT)
1510 NEXT JT
1520 NEXT IT
1530 REM
1540 REM Simulate genotypic effects for each trait
1550 REM
1560 FOR IT = 1 TO NT
1570 G(IT) = 0
1580 IF FF = 1 THEN 1640
1590 FOR IL = 1 TO NL
1600 X = RND
1610 IF X < F(IL,1) THEN G(IT) = G(IT) − A(IT,IL)
1620 IF X > F(IL,2) THEN G(IT) = G(IT) + A(IT,IL)
1630 NEXT IL
```

```
1640 NEXT IT
1650 REM
1660 REM Add simulated effects to get phenotypic values
1670 REM
1680 FOR IT = 1 TO NT
1690 PH(IT) = AV(IT) + G(IT) + B(IT) + W(IT)
1700 NEXT IT
1710 REM
1720 REM Write ID, IR, IP and phenotypic values to a file
1730 REM
1740 PRINT #1, USING "### ### ###"; ID; IR; IP;
1750 PRINT USING "### ### ###"; ID; IR; IP;
1760 FOR IT = 1 TO NT-1
1770 PRINT #1,USING " ####.##"; PH(IT);
1780 PRINT USING " ####.##"; PH(IT);
1790 NEXT IT
1800 PRINT #1,USING " ####.##"; PH(NT)
1810 PRINT USING " ####.##"; PH(NT)
1820 REM
1830 REM End simulation loops
1840 REM
1850 NEXT IP
1860 NEXT IR
1870 END
1880 REM Subroutine to generate NT independent N(0,1) deviates
1890 REM and return them in array R( ).
1900 REM
1910 REM Method based on polar coordinate method. See
1920 REM Knuth, D. E., The art of computer programming.
1930 REM Vol. 2. Seminumerical algorithms, Addison-Wesley,
1940 REM Reading, VA, 1969.
1950 REM
1960 FOR IT = 1 TO NT STEP 2
1970 U1 = 2*RND-1
1980 U2 = 2*RND-1
1990 S = U1*U1 + U2*U2
2000 IF S >= 1 THEN 1970
2010 S = SQR(-2*LOG(S)/S)
2020 R(IT) = S*U1
2030 IF IT = NT THEN 2050
2040 R(IT+1) = S*U2
2050 NEXT IT
2060 RETURN
```

APPENDIX 3:

PROGRAM FOR MULTINORMAL GENOTYPIC AND PHENOTYPIC VALUES

PROGRAM NAME: MNORM.BAS

LANGUAGE: BASIC

PURPOSE:
  This program generates genotypic and phenotypic values according to multivariate normal distributions, and stores the sample variances and covariances in a file.

DESCRIPTION:
  The program generates genotypic and phenotypic values that are distributed according to multivariate normal distributions by first using a polar transformation to convert two uniform numbers into two uncorrelated normal deviates with mean zero and variance of one. Then, the uncorrelated normal deviates are transformed into correlated variables by means of a multivariate transformation matrix. This process is repeated for each individual in the simulated sample. At the same time, appropriate sums, sums of squares, and sums of cross-products are accumulated for use in calculating the variances and covariances of the sample. While the methodology could be used for any multivariate normal data, the program is written to accept parameters which are pertinent only in the plant breeding context.

USER INSTRUCTIONS:
  When the program is loaded and run, the user will be prompted to indicate the number of traits and the size of the sample to be simulated. For each trait, the user will then specify the genotypic variance, heritability, and mean value. The user must also specify the genotypic and environmental correlations for each pair of traits. After computing the necessary transformation matrices, the program will prompt for an integer number to seed the random number generator. The user will also need to specify the name of a file into which the sample variances and covariances are to be written. The covariances are also printed on a printer as they are written into a file.
  The variances and covariances are written into the file in the form of a lower triangular matrix. That is, if $G(1,1)$ represents the genotypic variance for trait 1, and $G(1,2)$ represents the genotypic covariance between traits 1 and 2, then the order in the file will be $G(1,1),P(1,1),G(2,1),P(2,1),G(2,2),P(2,2),G(3,1),P(3,1)$, etc. Note that no identification is included in the diskette file.
  If the user wishes to use parts of this program for general simulation of multivariate normal populations, lines 580 through 800 provide a transformation from uncorrelated normal deviates with mean zero and variance one to correlated normal deviates with specified variances and covariances. To obtain one transformation matrix. say CG, one would remove all statements that refer to CE and TEMP2. The subroutine in lines 1700 and 1770 must be used to generate the independent normal deviates, and the coding in lines 1130 to 1190 would be required to simulate the correlated genotypic (GT) or environmental (ET) values.

PROGRAM LISTING:
  10 REM Program for simulating genotypic and phenotypic
  20 REM covariance matrices assuming a multinormal distribution.
  30 REM
  40 REM Covariance matrices are written to a file as well as
  50 REM to the screen.

```
 60 REM
 70 REM Clear screen and turn off key definitions
 80 REM
 90 CLS:KEY OFF
100 PRINT "Program for simulating multivariate normal distributions"
110 PRINT "= = = = = = = = = = = = = = = = = = = = = = = = = = = = = = =
         = = = = = = = = = = = = = = = = = = = = = = = ="
120 PRINT
130 REM
140 REM Prompt for interactive entry of data. Write to printer.
150 REM
160 INPUT "How many variables "; NV
170 DIM P(NV,NV),G(NV,NV),VG(NV,NV),CG(NV,NV),VE(NV,NV),CE(NV,NV),
    AVE(NV)
180 DIM GT(NV),ET(NV),X(NV),Y(NV),SP(NV),SG(NV)
190 INPUT "How many samples       "; NS
200 PRINT
210 FOR IV = 1 TO NV
220 PRINT "      For trait "; IV;" indicate the following:"
230 INPUT "        What is the genotypic variance "; VG(IV,IV)
240 LPRINT " For trait "; IV;", the genotypic variance is "; VG(IV,IV)
250 INPUT "        What is the heritability      "; H
260 LPRINT "      and the heritability is      "; H
270 VE(IV,IV) = VG(IV,IV)*(1 − H)/H
280 INPUT "        What is the average value      "; AVE(IV)
290 LPRINT "      and the average is       "; AVE(IV):LPRINT
300 NEXT IV
310 PRINT
320 FOR IV = 2 TO NV
330 FOR JV = 1 TO IV − 1
340 PRINT "        For trait "; JV;" and "; IV;" indicate the following:"
350 INPUT "        What is the genotypic correlation "; R
360 LPRINT "        Rg("; JV;","; IV;") = "; R
370 VG(IV,JV) = R * SQR(VG(IV,IV)*VG(JV,JV))
380 VG(JV,IV) = VG(IV,JV)
390 INPUT "        What is the environmental correlation "; R
400 LPRINT "        Re("; JV;","; IV;") = "; R
410 VE(IV,JV) = R * SQR(VE(IV,IV)*VE(JV,JV))
420 VE(JV,IV) = VE(IV,JV)
430 NEXT JV
440 NEXT IV
450 PRINT
460 PRINT " Please wait"
470 PRINT
480 REM
490 REM Transformation for multivariate normal distributions
500 REM See Naylor, T. H., Balintfy, J. L., Burdick, D. S.,
510 REM and Chu, K., Computer simulation techniques,
520 REM John Wiley and Sons, Inc., New York, NY, 1966.
530 REM
540 FOR IV = 1 TO NV
```

```
550 CG(IV,1) = VG(IV,1)/SQR(VG(1,1))
560 CE(IV,1) = VE(IV,1)/SQR(VE(1,1))
570 NEXT IV
580 FOR IV = 2 TO NV
590 TEMP = VG(IV,IV)
600 TEMP2 = VE(IV,IV)
610 FOR KV = 1 TO IV − 1
620 TEMP = TEMP − CG(IV,KV)^2
630 TEMP2 = TEMP2 − CE(IV,KV)^2
640 NEXT KV
650 CG(IV,IV) = SQR(TEMP)
660 CE(IV,IV) = SQR(TEMP2)
670 NEXT IV
680 IF NV < 3 THEN 810
690 FOR IV = 3 TO NV
700 FOR JV = 2 TO IV − 1
710 TEMP = VG(IV,JV)
720 TEMP2 = VE(IV,JV)
730 FOR KV = 1 TO JV − 1
740 TEMP = TEMP − CG(IV,KV)*CG(JV,KV)
750 TEMP2 = TEMP2 − CE(IV,KV)*CE(JV,KV)
760 NEXT KV
770 CG(IV,JV) = TEMP/CG(JV,JV)
780 CE(IV,JV) = TEMP2/CE(JV,JV)
790 NEXT JV
800 NEXT IV
810 CLS
820 REM
830 REM Prompt for random number seed and file in which to
840 REM store simulated covariance matrices.
850 REM
860 PRINT "Enter an integer value ( − 32768 to 32767) as seed"
870 INPUT "for the random number generator "; S%
880 PRINT
890 INPUT "What is name of file for output "; FO$
900 OPEN FO$ FOR OUTPUT AS #1
910 RANDOMIZE S%
920 REM
930 REM Zero covariance matrices
940 REM
950 FOR IV = 1 TO NV
960 FOR JV = 1 TO NV
970 P(IV,JV) = 0
980 G(IV,JV) = 0
990 NEXT JV
1000 NEXT IV
1010 REM
1020 REM Start to generate data.
1030 REM Print sample number to show program is running.
1040 REM Note: Locate command may not work on some computers.
1050 REM
```

```
1060 FOR IS = 1 TO NS
1070 LOCATE 10,12:PRINT IS
1080 FOR IV = 1 TO NV
1090 GOSUB 1700
1100 X(IV) = V1
1110 Y(IV) = V2
1120 NEXT IV
1130 FOR IV = 1 TO NV
1140 GT(IV) = 0
1150 ET(IV) = 0
1160 FOR JV = 1 TO NV
1170 GT(IV) = GT(IV) + CG(IV,JV)*X(JV)
1180 ET(IV) = ET(IV) + CE(IV,JV)*Y(JV)
1190 NEXT JV
1200 ET(IV) = ET(IV) + GT(IV)
1210 NEXT IV
1220 REM
1230 REM If user wishes, program could be modified at this
1240 REM point to have simulated genotypic and phenotypic
1250 REM values of each individual printed to a printer
1260 REM or to a file. GT(IV) and ET(IV) hold the genotypic
1270 REM and phenotypic values. AVE(IV) must be added to each.
1280 REM
1290 FOR IV = 1 TO NV
1300 SG(IV) = SG(IV) + GT(IV)
1310 SP(IV) = SP(IV) + ET(IV)
1320 FOR JV = 1 TO NV
1330 G(IV,JV) = G(IV,JV) + GT(IV)*GT(JV)
1340 P(IV,JV) = P(IV,JV) + ET(IV)*ET(JV)
1350 NEXT JV
1360 NEXT IV
1370 NEXT IS
1380 FOR IV = 1 TO NV
1390 FOR JV = 1 TO NV
1400 G(IV,JV) = (G(IV,JV) − SG(IV)/NS*SG(JV))/(NS − 1)
1410 P(IV,JV) = (P(IV,JV) − SP(IV)/NS*SP(JV))/(NS − 1)
1420 NEXT JV
1430 NEXT IV
1440 REM
1450 REM Print covariances of simulated values on a printer
1460 REM
1470 LPRINT "Genotypic and phenotypic variances and covariances "
1480 LPRINT:LPRINT" Sample size = "; NS
1490 LPRINT
1500 FOR IV = 1 TO NV
1510 FOR JV = 1 TO NV
1520 LPRINT USING " ##
     ##      ##.####^^^^      ##.####^^^^";IV,JV,G(IV,JV),P(IV,JV)
1530 NEXT JV
1540 NEXT IV
1550 REM
```

```
1560 REM Print matrices into a file (lower triangle mode)
1570 REM
1580 FOR JV = 1 TO NV
1590 FOR KV = 1 TO JV
1600 PRINT #1,G(JV,KV),P(JV,KV)
1610 NEXT KV
1620 NEXT JV
1630 END
1640 REM
1650 REM Subroutine to generate pair of random normal deviates.
1660 REM Use polar method of Knuth, D. E., The art of computer
1670 REM programming. Vol. 2., Seminumerical algorithms,
1680 REM Addison-Wesley, Reading, VA, 1969.
1690 REM
1700 V1 = 2*RND − 1
1710 V2 = 2*RND − 1
1720 S = V1*V1 + V2*V2
1730 IF S > = 1 THEN 1700
1740 S = SQR(− 2 *LOG(S)/S)
1750 V1 = V1*S
1760 V2 = V2*S
1770 RETURN
```

APPENDIX 4:

PROGRAM FOR CALCULATING MEAN SQUARES AND MEAN CROSS-PRODUCTS IN BALANCED FACTORIAL DESIGNS

PROGRAM NAME: BACOVA.FOR

LANGUAGE: FORTRAN

PURPOSE:

This FORTRAN program is designed to calculate mean squares and mean cross-products for all sources of variation in any balanced factorial design with up to 10 factors and with up to 10 variables measured on each experimental unit.

DESCRIPTION:

Written in FORTRAN, this program is designed to accept input for, and print results from, the IMSL subroutine AGBACP. At the same time that mean squares are calculated, the program loops through each pair of variables and calculates the mean cross-products through the relationship $xy = [(x + y)^2 - x^2 - y^2]/2$. The program prints mean squares and mean cross-products for each item in the model.

USER INSTRUCTIONS:

This program must be compiled and linked with the IMSL subroutine AGBACP. It cannot be used if that subroutine is not available. If available, the subroutine can be linked with this program and the executable module saved for subsequent analyses.

To use the program, data must first be stored in a file consisting of one record for each experimental unit. Each record must include the levels of each factor in the model (numbered from 1 to number of levels) as well as the data for each variable. Records need not be in order since the program will place the data in the proper order for analysis by AGBACP.

When the program is run, the user will be prompted to indicate whether or not instructions should be printed. The instructions, if printed, will indicate the format of the data file, as well as the method for specifying the model to be used in the analysis. Then, the user will be prompted to indicate how many factors there are in the experiment, how many levels there are of each factor, how many variables are to be analyzed, and the model to be used in the analysis.

The model must be specified as indicated for the instructions for using IMSL subroutine AGBACP (See Chapter 10 for a discussion). The model can vary from that of a simple randomized block design to complex designs involving crossed and nested factors. The program will handle most balanced designs but cannot be used if there are missing values.

BACOVA was written in Digital Equipment Systems FORTRAN for use on a DEC 2060 computer. To convert to other computers, the user will need to pay particular attention to the methods used for opening (OPEN) and closing (CLOSE) disk files, as well as for writing data to the screen (TYPE) and reading data from the keyboard (ACCEPT). If the word length for double precision alphanumeric variables is less than ten, it will be necessary to dimension the variable FNAME and to change the format in statement 110. If single precision alphanumeric variables are less then four characters long, similar changes will be required for the variable FMT and the format in statement 112.

PROGRAM LISTING:

```
DIMENSION IA(300),IB(10),FMT(20),ITL(40),ITR(40)
DIMENSION NL(10),ML(10),IWK(90),LST(1024),LOC(40)
```

```
       DIMENSION SS(41),NDF(41),Y(10000),X(3000,10)
       DIMENSION XIN(10),INF(10),ICHK(3000),SCP(40,55)
       DOUBLE PRECISION FNAME
       TYPE 100
100    FORMAT(/////' BALANCED ANALYSIS OF VARIANCE AND COVARIANCE')
       TYPE 101
101    FORMAT(' Do you want instructions (1 = yes,0 = no) '$)
       ACCEPT *,ISW
       IF(ISW.EQ.1)CALL INSTR
       TYPE 102
102    FORMAT(/' How many factors (2 to 10) '$)
       ACCEPT *,NF
       IF(NF.LT.2.OR.NF.GT.10)STOP
       NY = 1
       DO 1 IF = 1,NF
       TYPE 103,IF
103    FORMAT(' How many levels for factor ',I2,' : '$)
       ACCEPT *,NL(IF)
  1    NY = NY*(NL(IF)+1)
       NY = NY + 2**NF
       IF(NY.LE.10000)GO TO 2
       TYPE 104
104    FORMAT(/' Program cannot handle this size of data set ')
       STOP
  2    TYPE 105
105    FORMAT(/' How many variables to be analyzed (<11) '$)
       ACCEPT *,NV
       IF(NV.GT.10)STOP
       TYPE 106
106    FORMAT(/' Enter model (<41 terms; <301 characters; CAPS only)'/)
       ACCEPT 107,IA
107    FORMAT(300A1)
       ML(NF) = 1
       JJ = NF - 1
       DO 3 J = JJ,1,-1
       J1 = J + 1
  3    ML(J) = ML(J1)*NL(J1)
       NT = 1
       ITL(1) = 1
       DO 4 I = 1,300
       IF(IA(I).EQ.'(')ITR(NT)=I-1
       IF(IA(I).EQ.')')NT=NT+1
  4    IF(IA(I).EQ.')')ITL(NT)=I+1
       NT = NT-1
       IF(NT.LE.40)GO TO 5
       TYPE 108
108    FORMAT(' Error — too many terms in model')
       STOP
  5    TYPE 109
109    FORMAT(/' Enter name of data file (<11 characters) '$)
       ACCEPT 110,FNAME
```

```
110   FORMAT(A10)
      TYPE 111
111   FORMAT(/' Enter FORTRAN format (with brackets) '$)
      ACCEPT 112,FMT
112   FORMAT(20A4)
      NUM = 1
      DO 6 IF = 1,NF
  6   NUM = NUM*NL(IF)
      OPEN(UNIT = 21,FILE = FNAME)
      DO 10 J = 1,NUM
      READ(21,FMT)(INF(J),J = 1,NF),(XIN(J),J = 1,NV)
      II = 1
      DO 7 J = 1,NF
  7   II = II + (INF(J) − 1)*ML(J)
      ICHK(II) = 1
      DO 8 J = 1,NV
  8   X(II,J) = XIN(J)
 10   CONTINUE
      ICHECK = 0
      DO 11 I = 1,NUM
 11   IF(ICHK(I).NE.1)ICHECK = 1
      IF(ICHECK.EQ.0)GO TO 12
      TYPE 113
113   FORMAT(/' Error in factor levels — check data ')
      STOP
 12   CONTINUE
      DO 16 J = 1,NV
      II = 2**NF
      DO 13 I = 1,NUM
      II = II + 1
 13   Y(II) = X(I,J)
      CALL AGBACP(NF,NL,IA,Y,IWK,LST,LOC,SS,NDF,IER)
      IF(IER.EQ.0)GO TO 14
      TYPE 114
114   FORMAT(/' Error detected in AGBACP subroutine ')
      IF(IER.EQ.131)TYPE 115
115   FORMAT(' Error in model — check instructions')
      TYPE 207,IA
207   FORMAT(1X,79A1)
      STOP
 14   CONTINUE
      DO 15 IT = 1,NT
 15   SCP(IT,J) = SS(IT)
 16   CONTINUE
      JK = NV
      DO 20 J = 1,NV
      DO 20 K = 1,J
      IF(J.EQ.K)GO TO 20
      JK = JK + 1
      II = 2**NF
      DO 17 I = 1,NUM
      II = II + 1
```

```
   17   Y(II) = X(I,J) + X(I,K)
        CALL AGBACP(NF,NL,IA,Y,IWK,LST,LOC,SS,NDF,IER)
        DO 18 IT = 1,NT
   18   SCP(IT,JK) = (SS(IT) − SCP(IT,J) − SCP(IT,K))/2.0
   20   CONTINUE
        TYPE 116
  116   FORMAT(///' Analysis of variance and covariance'/)
        TYPE 117
  117   FORMAT(' Source          df        Mean squares and cross-products')
        NVV = NV*(NV+1)/2
        DO 21 IT = 1,NT
        DO 21 J=1,NVV
   21   SCP(IT,J) = SCP(IT,J)/NDF(IT)
        DO 30 IT=1,NT
        L1 = ITL(IT)
        L2 = ITR(IT)
        I = 0
        DO 22 J = L1,L2
        I = I + 1
        IB(I) = IA(J)
        II = L2 − L1 + 2
        DO 23 J = II,10
        IB(J) = ' '
        TYPE 118,IB,NDF(IT),SCP(IT,1)
  118   FORMAT(/lX,10A1,I5,1X,G12.5)
        IF(NV.LE.1)GO TO 30
        JV = NV + 1
        DO 25 IV = 2,NV
        JV = JV + IV − 2
        KV = JV + IV − 2
   25   TYPE 119,(SCP(IT,J),J=JV,KV),SCP(IT,IV)
  119   FORMAT(17X,FG12.5/18X,5G12.5)
   30   CONTINUE
        TYPE 120
  120   FORMAT(/
        ' = = = = = = = = = = = = = = = = = = = = = = = = = = = = = = = =',
    1   ' = = = = = = = = = = = = = = = = = = = = = = = = = = = = = = = =
        = = = = '//////)
        CLOSE(UNIT = 21)
        END
        SUBROUTINE INSTR
        TYPE 1
    1   FORMAT(' To use this program, you must specify:')
        TYPE 2
    2   FORMAT(' a) a model in the form A(I)B(J)AB(IJ)E(IJK),')
        TYPE 3
    3   FORMAT(' b) levels of each factor in the model,')
        TYPE 4
    4   FORMAT(' c) number of variables to be analyzed,')
        TYPE 5
    5   FORMAT(' d) name of data file, and')
        TYPE 6
```

```
 6  FORMAT(' e) format of the data file.')
    TYPE 7
 7  FORMAT(//' In the model A(I)B(J)AB(IJ)C(IJK)E(IJKL), there are')
    TYPE 8
 8  FORMAT(' five terms A(I), B(J), AB(IJ), C(IJK) and E(IJKL).')
    TYPE 9
 9  FORMAT(' The brackets enclose the subscripts of each term.')
    TYPE 10
10  FORMAT(' For each term, the associated subscripts must be the')
    TYPE 11
11  FORMAT(' last in the list for that term. For example, the')
    TYPE 12
12  FORMAT(' subscript K in C(IJK) is associated with the effect C;')
    TYPE 13
13  FORMAT(' the I and J indicate that effect C is nested within')
    TYPE 14
14  FORMAT(' effects A(I) and B(J).')
    TYPE 15
15  FORMAT(/'        E must be the last term in the model and it must')
    TYPE 16
16  FORMAT(' include the full set of subscripts — one for each')
    TYPE 17
17  FORMAT(' factor in the model.')
    TYPE 18
18  FORMAT('         Press return to continue')
    ACCEPT 100,KEY
100 FORMAT(A1)
    TYPE 19
19  FORMAT(//' Each record in the data file must contain:')
    TYPE 20
20  FORMAT(' level of factor 1 (e.g. A) from 1 to #levels of 1,')
    TYPE 21
21  FORMAT(' level of factor 2 (e.g. B) from 1 to #levels of 2,')
    TYPE 22
22  FORMAT('      .       .       .       .      .      .      .      .')
    TYPE 23
23  FORMAT(' level of factor n from 1 to #levels for n,')
    TYPE 24
24  FORMAT(' value for variable 1,')
    TYPE 25
25  FORMAT('        .      .      .        .')
    TYPE 26
26  FORMAT(' value for variable m.')
    TYPE 27
27  FORMAT(/' The format must specify the factors levels in')
    TYPE 28
28  FORMAT(' integer (Iw) format and the data in floating')
    TYPE 29
29  FORMAT(' (Fw.d) format. The records do not have to be in')
    TYPE 30
```

```
30  FORMAT(' any specific order.')
    TYPE 18
    ACCEPT 100,KEY
    RETURN
    END
```

PROGRAM FOR CALCULATING SPECIFIED MEAN SQUARES AND MEAN CROSS-PRODUCTS

PROGRAM NAME: BACOVA1.BAS

LANGUAGE: BASIC

PURPOSE:

This program is designed to calculate mean squares and mean cross-products for specified terms in balanced factorial designs.

DESCRIPTION:

For any term in the model for a particular analysis, the user specifies whether the subscript corresponding to each factor is associated with an effect in the term, floating, or not in the term. The data are then read from a diskette file and the means, over all factors whose corresponding subscripts are not included in the term, are calculated. The second step in the algorithm is to cycle through the associated subscripts and to calculate the differences between new means and those resulting from the previous cycle. The final step is to square or multiply appropriate effects, sum, and multiply by proper coefficients to get the mean squares and cross-products for the specified term. The name of the term, its degrees of freedom, and the mean squares and cross-products are then written into an output file.

This program permits analysis of the largest possible experiment when random access memory is limited. The data must be read from the input file each time a new term is designated. It is not necessary to calculate mean squares and cross-products for all the terms in the model if only a portion of them are required for developing a selection index. The algorithm used in this program is described in greater detail in Chapter 10.

USER INSTRUCTIONS:

When this program is run, the user will be prompted to give the name of the file that contains the data. The file must contain the data for each experimental unit in the following order: level of factor 1, level of factor 2, . . . , level of factor n, data for variable 1, . . . , data for variable m. Factor levels must be represented by consecutive integer values from 1 to the number of levels for the particular factor.

The user must also specify the name of a file into which the calculated mean squares and cross-products can be written. The program will then prompt for number of factors, number of variables, and number of levels of each factor.

For each term, the user must specify whether or not each subscript is associated with an effect in the term, is a floating subscript, or does not appear in the term. See Chapter 10 for a detailed discussion of associated and floating subscripts. The program will then read the data, make the necessary calculations, print the mean squares and cross-products on the screen and write them to the output file, and prompt to see if another term is to be calculated.

In some cases, it may be desirable to skip a field in the data file. For example, if analyzing data in a file created by GENSIM.BAS (Appendix 2), it would be desirable to skip the first field in each record since it is the identification number for the population. This can be facilitated by adding a statement such as

865 IF IFACT = 1 THEN INPUT #1,IL(IFACT)

prior to running the program. A similar statement at line 895 would allow the user to skip data for one variable.

PROGRAM LISTING:

```
 10 REM Program for computing mean squares and cross products.
 20 REM Method based on modification of algorithm described by
 30 REM Hemmerle, W. J., Statistical computations on a digital
 40 REM computer, Blaisdell Publishing Co., Waltham, MA, 1967.
 50 REM
 60 REM Clear screen, set base option, print instructions.
 70 REM
 80 CLS:OPTION BASE 1:KEY OFF
 90 GOSUB 1960
100 REM
110 REM Prompt for file names, number of factors and factor levels.
120 REM
130 CLS
140 INPUT " Enter name of file containing data "; FIL$
150 INPUT " Enter name of file where mean squares and products to be saved "; FOUT$
160 INPUT " Enter number of factors "; NFACT
170 INPUT " Enter number of variables "; NVAR
180 DIM NL(NFACT),ASSOC(NFACT),FLOAT(NFACT),ML(NFACT),
    WORK(4000,NVAR)
190 DIM IL(NFACT),XDATA(NVAR)
200 NREC = 1
210 FOR IFACT = 1 TO NFACT
220 PRINT USING " Enter number of levels for factor ## "; IFACT;
230 INPUT NL(IFACT)
240 NREC = NREC * NL(IFACT)
250 NEXT IFACT
260 PRINT
270 REM
280 REM open output file
290 REM
300 OPEN FOUT$ FOR OUTPUT AS #2
310 REM
320 REM begin major loop
330 REM
340 INPUT "Enter name for this component "; MNAME$
350 DF = 1
360 NMEAN = 1
370 NASSOC = 0 : NFLOAT = 0 : NUM = 1
380 FOR IFACT = 1 TO NFACT
390 PRINT USING " For this component, is subscript for factor ## A, F or N "; IFACT;
400 INPUT R$
410 IF R$ <> "A" AND R$ <> "a" THEN 470
420 ASSOC(IFACT) = NL(IFACT)
430 FLOAT(IFACT) = 0
440 NASSOC = NASSOC + 1
450 NUM = NUM * NL(IFACT)
460 GOTO 590
470 IF R$ <> "F" AND R$ <> "f" THEN 530
480 ASSOC(IFACT) = 0
490 FLOAT(IFACT) = NL(IFACT)
```

```
500 NFLOAT = NFLOAT + 1
510 NUM = NUM * NL(IFACT)
520 GOTO 590
530 IF R$ <> "N" AND R$ <> "n" THEN 590
540 ASSOC(IFACT) = 0
550 FLOAT(IFACT) = 0
560 NMEAN = NMEAN * NL(IFACT)
570 GOTO 590
580 PRINT "Error in input — — — Redo!!!":GOTO 400
590 NEXT IFACT
600 REM
610 REM Read data into work — calculate means in process
620 REM
630 CLS
640 PRINT
650 PRINT "Reading data and calculating means ";
660 REM
670 FOR INUM = 1 TO NUM
680 FOR IVAR = 1 TO NVAR
690 WORK(INUM,IVAR) = 0
700 NEXT IVAR
710 NEXT INUM
720 FOR IFACT = 1 TO NFACT − 1
730 ML(IFACT) = 0
740 IF ASSOC(IFACT) = 0 AND FLOAT(IFACT) = 0 THEN 800
750 ML(IFACT) = 1
760 FOR JFACT = IFACT + 1 TO NFACT
770 IF ASSOC(JFACT) = 0 AND FLOAT(JFACT) = 0 THEN 790
780 ML(IFACT) = ML(IFACT)*NL(JFACT)
790 NEXT JFACT
800 NEXT IFACT
810 ML(NFACT) = 1
820 IF ASSOC(NFACT) = 0 AND FLOAT(NFACT) = 0 THEN ML(NFACT) = 0
830 OPEN FIL$ FOR INPUT AS #1
840 FOR IREC = 1 TO NREC
850 FOR IFACT = 1 TO NFACT
860 INPUT #1,IL(IFACT)
870 NEXT IFACT
880 FOR IVAR = 1 TO NVAR
890 INPUT #1,XDATA(IVAR)
900 NEXT IVAR
910 J1 = 1
920 FOR IFACT = 1 TO NFACT
930 IF ASSOC(IFACT) = 0 AND FLOAT(IFACT) = 0 THEN 990
940 J1 = J1 + (IL(IFACT) − 1)*ML(IFACT)
950 REM
960 REM Following statement to show program is running
970 REM
980 LOCATE ,50:PRINT J1;
990 NEXT IFACT
1000 FOR IVAR = 1 TO NVAR
```

```
1010 WORK(J1,IVAR) = WORK(J1,IVAR) + XDATA(IVAR)/NMEAN
1020 NEXT IVAR
1030 NEXT IREC
1040 REM End of data input
1050 REM
1060 REM Calculate degrees of freedom for this component
1070 REM
1080 DF = 1
1090 FOR IFACT = 1 TO NFACT
1100 IF ASSOC(IFACT) <> 0 THEN DF = DF * (ASSOC(IFACT) − 1)
1110 IF FLOAT(IFACT) <> 0 THEN DF = DF * FLOAT(IFACT)
1120 NEXT IFACT
1130 PRINT #2,MNAME$;";"; DF
1140 PRINT
1150 PRINT "Mean squares and cross-products for "; MNAME$
1160 PRINT "        with "; DF;" degrees of freedom"
1170 CLOSE #1
1180 REM
1190 REM Calculate deviations using modification of algorithm
1200 REM given in chapter 5 of Hemmerle.
1210 REM
1220 FOR LOOP = 1 TO NFACT
1230 IF ASSOC(LOOP) = 0 THEN 1710
1240 NPM = ASSOC(LOOP)
1250 MEANST = NUM/NPM
1260 INC1 = 1
1270 IF LOOP = NFACT THEN 1320
1280 FOR IFACT = NFACT TO LOOP + 1 STEP −1
1290 IF ASSOC(IFACT) <> 0 THEN INC1 = INC1 * ASSOC(IFACT)
1300 IF FLOAT(IFACT) <> 0 THEN INC1 = INC1 * FLOAT(IFACT)
1310 NEXT IFACT
1320 LOC1 = 1
1330 LOC2 = NUM + 1
1340 M = 1
1350 FOR I = 1 TO MEANST STEP INC1
1360 J2 = I + INC1 − 1
1370 FOR J = I TO J2
1380 L = M
1390 I1 = LOC2 + J − 1
1400 FOR IVAR = 1 TO NVAR
1410 WORK(I1,IVAR) = 0
1420 NEXT IVAR
1430 FOR K = 1 TO NPM
1440 I2 = LOC1 + L − 1
1450 FOR IVAR = 1 TO NVAR
1460 WORK(I1,IVAR) = WORK(I1,IVAR) + WORK(I2,IVAR)/NPM
1470 NEXT IVAR
1480 L = L + INC1
1490 NEXT K
1500 M = M + 1
1510 NEXT J
```

```
1520 M = L − INC1 + 1
1530 NEXT I
1540 M = 1
1550 FOR I = 1 TO MEANST STEP INC1
1560 J2 = I + INC1 − 1
1570 FOR J = I TO J2
1580 L = M
1590 I1 = LOC2 + J − 1
1600 FOR K = 1 TO NPM
1610 I2 = LOC1 + L − 1
1620 FOR IVAR = 1 TO NVAR
1630 WORK(I2,IVAR) = WORK(I2,IVAR) − WORK(I1,IVAR)
1640 NEXT IVAR
1650 L = L + INC1
1660 NEXT K
1670 M = M + 1
1680 NEXT J
1690 M = L − INC1 + 1
1700 NEXT I
1710 NEXT LOOP
1720 REM
1730 REM Accumulate sums of squares and cross-products
1740 REM
1750 FOR IVAR = 1 TO NVAR
1760 FOR JVAR = IVAR TO NVAR
1770 SUMSQR = 0
1780 FOR INUM = 1 TO NUM
1790 SUMSQR = SUMSQR + WORK(INUM,IVAR)*WORK(INUM,JVAR)
1800 NEXT INUM
1810 SUMSQR = SUMSQR/DF*NMEAN
1820 PRINT #2,SUMSQR
1830 PRINT USING " ## ## ######.#### "; IVAR,JVAR,SUMSQR
1840 NEXT JVAR
1850 NEXT IVAR
1860 REM
1870 REM End of computations
1880 REM See if another is to be done
1890 REM
1900 PRINT
1910 INPUT "Want to do another (Y)es or (N)o "; R$
1920 IF R$ = "Y" OR R$ = "y" THEN 330
1930 IF R$ <> "N" AND R$ <> "n" THEN 1910
1940 CLOSE #2
1950 END
1960 KEY OFF:CLS:LOCATE 1,10:PRINT"Program for calculating mean squares and"
1970 LOCATE 2,10:PRINT"mean cross-products in a balanced design"
1980 LOCATE 3,10:PRINT"= = = = = = = = = = = = = = = = = = = = = = = = = =
     = = = = = = = = = = = = = = ="
1990 LOCATE 5,1:PRINT"Data must be stored in a file in the following way"
2000 LOCATE 6,5:PRINT"The first items on each record must be the levels"
2010 LOCATE 7,5:PRINT"of each factor (from 1 to # levels)."
```

```
2020 LOCATE 8,5:PRINT"These are then followed by the data for each variable."
2030 LOCATE 10,1:PRINT"To use the program, you must understand the meaning"
2040 LOCATE 11,5:PRINT"of (A)ssociated and (F)loating subscripts as used by"
2050 LOCATE 12,5:PRINT"W. J. Hemmerle, Statistical computations on a"
2060 LOCATE 13,5:PRINT"        digital computer (QA276.H48)"
2070 LOCATE 14,5:PRINT"In the model Y(ijkl) = u + s(i) + t(ij) + p(k)"
2080 LOCATE 15,5:PRINT"        + sp(ik) + tp(ijk) + e(ijkl),"
2090 LOCATE 16,5:PRINT"component tp(ijk) has two Associated subscripts (j and k)"
2100 LOCATE 17,5:PRINT"and one Floating subscript (i)."
2110 LOCATE 19,1:PRINT"This program calculates mean squares and cross-products"
2120 LOCATE 20,1:PRINT"for one component at a time. Any component may be "
2130 LOCATE 21,1:PRINT"specified — it is not necessary to evaluate all terms!"
2140 LOCATE 22,1:PRINT"For each component, you will be asked to specify if each"
2150 LOCATE 23,1:PRINT"possible subscript is (A)ssociated,(F)loating or"
2160 LOCATE 24,1:PRINT"(N)ot part of the term.        Press any key to continue.";
2170 REM
2180 REM Wait for key press
2190 REM
2200 WHILE INKEY$ = "":WEND
2210 RETURN
```

PROGRAM FOR CALCULATING MEAN SQUARES AND MEAN CROSS-PRODUCTS FOR BALANCED DESIGNS

PROGRAM NAME: BACOVA2.BAS

LANGUAGE: BASIC

PURPOSE:

This program calculates and displays mean squares and mean cross-products for all terms in the model for a balanced design.

DESCRIPTION:

This program reads an input file and stores the data in a particular order in a work area. A series of orthogonal transformations are then used to replace the data for each variable by transformed data consisting of a correction factor, all main effects, and all possible interactions among the design factors. The sums of squares or products for the various effects are then collected according to the model specified by the user.

This program differs from BACOVA1 (Appendix 5) in that the user specifies the complete model and the program calculates means squares and cross-products corresponding to each term in the model.

USER INSTRUCTIONS:

When the program is loaded and run, it first prompts for the name of the file from which the data is to be read and the name of the file into which the calculated mean squares and cross-products are to be written. It then branches to a subroutine that prompts for and processes the design model. The model must be specified as outlined in Chapter 10. That is, effects are indicated by single letters. Subscripts, also single letters, must be included within parentheses. Subscripts associated with effects in a term must be last in the subscript list for that term. The last term in the model must be designated by the letter E and must be followed in brackets by a complete list of subscripts. For a two-way analysis of variance with subsamples, the model could be indicated by A(I)B(J)AB(IJ)E(IJK).

Once the program has determined which subscripts are associated with which effects, and which are floating, the user will be prompted to indicate the number of levels of each subscript, and the number of variables to be analyzed. The program will then proceed to calculate all mean squares and cross-products and write them to the output file.

As with BACOVA1 (Appendix 5), the user may wish to skip particular fields in the data file. This can be done by adding statements at line 405 to skip a field within the list of factor levels, or at line 495 to skip a field within the list of data items. For example, to skip the second factor level, use

$$405 \text{ IF J } = 2 \text{ THEN INPUT \#1,ML}$$

and, to skip the first variable, use

$$495 \text{ IF J } = 1 \text{ THEN INPUT \#1, Y(N1,J)}$$

PROGRAM LISTING:

```
10 REM Program for calculating mean squares and mean
20 REM cross-products for balanced designs
```

```
 30 REM
 40 REM Based on algorithm by Howell, J. R., Algorithm 359.
 50 REM Factorial analysis of variance, Comm. of the A.C.M.,12,631,1969.
 60 REM
 70 REM
 80 REM Clear screen, turn off key definitions, set base to 1.
 90 REM
100 CLS:KEY OFF:OPTION BASE 1
110 PRINT:PRINT"Mean squares and cross-products for balanced designs"
120 PRINT:PRINT
130 REM
140 REM Prompt for file names, call MODEL subroutine
150 REM
160 INPUT "What is the name of file containing data "; FI$
170 INPUT "What is the name of file in which ms and mcp to be stored "; FO$
180 PRINT
190 GOSUB 2150
200 NF = NS
210 DIM NL(NF)
220 FX=0
230 N = 1
240 FOR I = 1 TO NF
250 PRINT "How many levels for subscript "; S$(I);
260 INPUT NL(I)
270 IF NL(I) > FX THEN FX = NL(I)
280 N=N*NL(I)
290 NEXT I
300 PRINT
310 INPUT "How many variables "; NV
320 DIM Y(N,NV),Z(N),ROW(FX)
330 REM
340 REM Read data
350 REM
360 OPEN FI$ FOR INPUT AS #1
370 FOR I = 1 TO N
380 N1 = 1
390 FOR J = 1 TO NF
400 INPUT #1,ML
410 ML=ML-1
420 IF J = NF THEN 460
430 FOR K = J+1 TO NF
440 ML = ML*NL(K)
450 NEXT K
460 N1 = N1 + ML
470 NEXT J
480 FOR J = 1 TO NV
490 INPUT #1, Y(N1,J)
500 NEXT J
510 NEXT I
520 CLOSE 1
530 REM
```

```
540 REM Start major loop with algorithm 359
550 REM
560 REM Print variable, factor and level to let user know
570 REM program is working.
580 REM
590 LOCATE 20,1:PRINT "Variable =        "
600 LOCATE 21,1:PRINT "Factor =       "
610 LOCATE 22,1:PRINT "Level =       "
620 FOR J = 1 TO NV
630 LOCATE 20,12:PRINT J
640 FOR JF = 1 TO NF
650 LOCATE 21,12:PRINT JF
660 II = 1
670 KK = NF − JF + 1
680 NR = NL(KK)
690 FOR JJ = 1 TO NR
700 LOCATE 22,12:PRINT JJ
710 REM subroutine arow
720 IF JJ <> 1 THEN 790
730 AA = NR
740 EL = 1/SQR(AA)
750 FOR KK = 1 TO NR
760 ROW(KK) = EL
770 NEXT KK
780 GOTO 900
790 J1 = JJ − 1
800 RJ = JJ
810 AA = SQR(RJ * RJ − RJ)
820 EL = 1 /AA
830 FOR KK = 1 TO J1
840 ROW(KK) = EL
850 NEXT KK
860 FOR KK = JJ TO NR
870 ROW(KK) = 0
880 NEXT KK
890 ROW(JJ) = (1 − RJ)/AA
900 FOR KK = 1 TO N STEP NR
910 Z(II) = 0
920 FOR LL = 1 TO NR
930 K1 = KK + LL − 1
940 Z(II) = Z(II) + ROW(LL)*Y(K1,J)
950 NEXT LL
960 II = II + 1
970 NEXT KK
980 NEXT JJ
990 FOR JJ = 1 TO N
1000 Y(JJ,J) = Z(JJ)
1010 NEXT JJ
1020 NEXT JF
1030 NEXT J
1040 REM
```

```
1050 REM Effects calculated, now SS and SCP
1060 REM
1070 D1 = 2^NF:D2 = (NF + 1)*NF/2
1080 DIM SS(D1,D2),CO(NF),DF(D1)
1090 FOR J = 1 TO D1
1100 DF(J) = 0
1110 NEXT J
1120 FOR J = 1 TO NF
1130 CO(J) = 0
1140 NEXT J
1150 FOR I = 1 TO N
1160 FOR J = NF TO 2 STEP − 1
1170 IF CO(J) < NL(J) THEN 1200
1180 CO(J) = 0
1190 CO(J − 1) = CO(J − 1) + 1
1200 NEXT J
1210 K = 1
1220 FOR J = 1 TO NF
1230 IF CO(J) > 0 THEN K = K + 2^(NF − J)
1240 NEXT J
1250 DF(K) = DF(K) + 1
1260 JK = 0
1270 FOR J = 1 TO NV
1280 FOR JJ = 1 TO J
1290 JK = JK + 1
1300 SS(K,JK) = SS(K,JK) + Y(I,J)*Y(I,JJ)
1310 NEXT JJ
1320 NEXT J
1330 CO(NF) = CO(NF) + 1
1340 NEXT I
1350 REM
1360 REM Print mean squares and cross-products
1370 REM
1380 OPEN FO$ FOR OUTPUT AS #2
1390 CLS
1400 PRINT
1410 PRINT"Source df Mean squares and cross-products"
1420 PRINT"= = = = = = = = = = = = = = = = = = = = = = = = = = = = = = = = =
     = = = = = = = = = ="
1430 FOR IT = 1 TO NT
1440 PRINT T$(IT);
1450 PRINT #2,T$(IT)
1460 IDF = 0
1470 JK = 0
1480 FOR J = 1 TO NV
1490 FOR JJ = 1 TO J
1500 JK = JK + 1
1510 SS(1,JK) = 0
1520 NEXT JJ
1530 NEXT J
1540 FOR I = 2 TO 2^NF
```

```
1550 IF ((I − 1) OR SUB(IT)) <> SUB(IT) THEN 1660
1560 IF ((I − 1) AND SYM(IT)) <> SYM(IT) THEN 1660
1570 JK = 0
1580 FOR J = 1 TO NV
1590 FOR JJ = 1 TO J
1600 JK = JK + 1
1610 SS(1,JK) = SS(1,JK) + SS(I,JK)
1620 NEXT JJ
1630 NEXT J
1640 IDF = IDF + DF(I)
1650 DF(I) = 0
1660 NEXT I
1670 PRINT TAB(10 − LEN(STR$(IDF)) + 1); IDF;
1680 PRINT #2,IDF
1690 JK = 0
1700 FOR J = 1 TO NV
1710 FOR JJ = 1 TO J
1720 JK = JK + 1
1730 PRINT USING " ####.#^^^";SS(1,JK)/IDF;
1740 PRINT #2,SS(1,JK)/IDF
1750 NEXT JJ
1760 PRINT TAB(12);
1770 NEXT J
1780 PRINT
1790 NEXT IT
1800 JK = 0
1810 FOR J = 1 TO NV
1820 FOR JJ = 1 TO J
1830 JK = JK + 1
1840 SS(1,JK) = 0
1850 NEXT JJ
1860 NEXT J
1870 IDF=0
1880 FOR I = 2 TO 2^NF
1890 IF DF(I) = O THEN 1980
1900 JK = 0
1910 FOR J = 1 TO NV
1920 FOR JJ = 1 TO J
1930 JK = JK + 1
1940 SS(1,JK) = SS(1,JK) + SS(I,JK)
1950 NEXT JJ
1960 NEXT J
1970 IDF = IDF + DF(I)
1980 NEXT I
1990 PRINT "Error"; TAB(10 − LEN(STR$(IDF)) + 1); IDF;
2000 PRINT #2,"Error"
2010 PRINT #2,IDF
2020 JK = 0
2030 FOR J = 1 TO NV
2040 FOR JJ = 1 TO J
2050 JK = JK + 1
```

```
2060 PRINT USING " ####.#^^^^";SS(1,JK)/IDF;
2070 PRINT #2,SS(1,JK)/IDF
2080 NEXT JJ
2090 PRINT TAB(12);
2100 NEXT J
2110 PRINT
2120 CLOSE 2
2130 END
2140 REM
2150 REM Accept and process model for Balanced Analysis of Covariance
2160 REM
2170 INPUT "Enter model :"; M$
2180 REM
2190 REM Determine number of terms (NT)
2200 REM
2210 NT = 0
2220 FOR J = 1 TO LEN(M$)
2230 IF MID$(M$,J,1) = "(" THEN NT = NT + 1
2240 NEXT J
2250 REM
2260 REM Determine number of subscripts (NS)
2270 REM
2280 NS = 0:SW = 0
2290 FOR J = 1 TO LEN(M$)
2300 IF MID$(M$,J,1) = "E" OR MID$(M$,J,1) = "e" THEN SW = 1:GOTO 2330
2310 IF SW = 0 THEN 2330
2320 IF MID$(M$,J,1) <> "(" AND MID$(M$,J,1) <> ")" THEN NS = NS + 1
2330 NEXT J
2340 REM
2350 REM Determine effects and subscripts
2360 REM
2370 DIM E$(NS),S$(NS),SYM(NT),SUB(NT),T$(NT)
2380 SW = 0:NE = 0:MS = 0:ISYM = 1:ISUB = 1
2390 FOR J = 1 TO LEN(M$)
2400 C$ = MID$(M$,J,1)
2410 IF C$ = "(" THEN SW = 1
2420 IF C$ = "(" THEN ISYM = ISYM + 1
2430 IF C$ = ")" THEN SW = 0
2440 IF C$ = ")" THEN ISUB = ISUB + 1
2450 IF C$ = "(" OR C$ = ")" THEN 2660
2460 IF C$ = "e" OR C$ = "E" THEN 2660
2470 IF SW = 1 THEN 2580
2480 T$(ISYM) = T$(ISYM)+C$
2490 ES = 0
2500 FOR IE = 1 TO NE
2510 IF E$(IE) = C$ THEN ES = 1
2520 NEXT IE
2530 IF ES = 0 THEN NE = NE + 1:E$(NE) = C$
2540 FOR IE = 1 TO NE
2550 IF E$(IE) = C$ THEN SYM(ISYM) = SYM(ISYM) OR (2^(NS-IE))
2560 NEXT IE
```

```
2570 GOTO 2660
2580 SS = 0
2590 FOR IS = 1 TO MS
2600 IF S$(IS) = C$ THEN SS = 1
2610 NEXT IS
2620 IF SS = 0 THEN MS = MS + 1:S$(MS) = C$
2630 FOR IS = 1 TO MS
2640 IF S$(IS) = C$ THEN SUB(ISUB) = SUB(ISUB) OR (2^(NS-IS))
2650 NEXT IS
2660 NEXT J
2670 NT = NT - 1
2680 RETURN
```

APPENDIX 7:

PROGRAM FOR CALCULATING EXPECTATIONS OF MEAN SQUARES

PROGRAM NAME: EMS.BAS

LANGUAGE: BASIC

PURPOSE:
This program calculates and prints the expectations of mean squares for any balanced design.

DESCRIPTION:
The design model, specified in computer readable form, is analyzed to determine which subscripts are associated with which design effects. Once the user has specified the number of levels of each factor and whether or not that factor represents a random or a fixed effect, the program calculates expectations of mean squares for each term in the model. The rules used in developing the expectations are those given by C. R. Hicks.

USER INSTRUCTIONS:
When the program is loaded and run, the user must first specify the model for the analysis in a computer readable format. The required format is discussed in Chapter 10 and Appendices 4 and 6. The following examples provide further guidance.

1.	one-way design	A(I)E(IJ)
2.	two-way design	A(I)B(J)E(IJ)
3.	two-way design with subsamples	A(I)B(J)AB(IJ)E(IJK)
4.	split-plot in time	R(I)A(J)RA(IJ)B(K)RB(IK)
		AB(JK)E(IJK)
5.	two factor factorial in a one-way design with subsampling	A(I)B(J)AB(IJ)S(IJK)E(IJKL)

Once the model has been entered, the user will then be asked to indicate whether each subscript is associated with a random or fixed variable as well as how many levels there are of each factor. The program will then print a table consisting of the source of variation (the term in the model) and its expected mean square. In the expectation of mean squares, the symbol $ is used to denote a variance component while the symbol # is used to indicate the average square of fixed effects. The user may wish to change these to some other value by changing statements 620 and 650 as well as statements 1030 and 1040.

PROGRAM LISTING:
```
10 REM
20 REM Program to calculate expectations of mean squares.
30 REM Based on rules given by Hicks, C. R., Fundamental
40 REM concepts in the design of experiments, Holt,
50 REM Reinhart and Winston, New York, NY, 1973.
60 REM
70 REM Clear screen, turn key definitions off, call MODEL.
80 REM
90 KEY OFF:CLS
100 PRINT" Expectations of means squares for balanced designs"
```

```
110 PRINT
120 GOSUB 1070
130 DIM LV(NS),TB(NT,NS),FR$(NS),SY$(NT),CV(NS)
140 REM
150 REM Determine levels of each factor and whether it is
160 REM a fixed or random effect.
170 REM
180 FOR IS = 1 TO NS
190 PRINT "For this model, subscript "; S$(IS)
200 PRINT $        is associated with effect "; E$(IS)
210 INPUT "        How many levels for this subscript "; LV(IS)
220 INPUT "        Is this subscript random (R) or fixed (F) "; FR$(IS)
230 IF FR$(IS) = "r" THEN FR$(IS) = "R"
240 IF FR$(IS) = "f" THEN FR$(IS) = "F"
250 IF FR$(IS) <> "F" AND FR$(IS) <> "R" THEN 220
260 NEXT IS
270 REM
280 REM Apply Hick's rule 2, page 177.
290 REM
300 FOR IT = 1 TO NT
310 FOR IS = 1 TO NS
320 IF SUB(IT) AND 2^(NS − IS) THEN 340
330 TB(IT,IS) = LV(IS)
340 NEXT IS
350 NEXT IT
360 REM
370 REM Apply Hicks' rule 3, page 178
380 REM
390 FOR IT = 1 TO NT
400 FOR IS = 1 TO NS
410 IF TB(IT,IS) <> 0 THEN 440
420 A = 2^(NS − IS) AND SYM(IT):IF A THEN GOTO 440
430 IF TB(IT,IS) = 0 THEN TB(IT,IS) = 1
440 NEXT IS
450 NEXT IT
460 REM
470 REM Apply Hicks' rule 4, Page 178
480 REM
490 FOR IT = 1 TO NT
500 FOR IS = 1 TO NS
510 IF TB(IT,IS) <> 0 THEN 530
520 IF FR$(IS) = "R" THEN TB(IT,IS) = 1
530 NEXT IS
540 NEXT IT
550 REM
560 REM Hicks' rule 5 covered in initial zeroing.
570 REM Apply Hicks' rule 6, page 179.
580 REM Print expectations using $ for variance component
590 REM and # for average square of fixed effects.
600 REM
610 FOR IT = 1 TO NT
```

```
620 R$ = "#"
630 FOR J = 1 TO NS
640 FOR IS = 1 TO NS
650 IF (SYM(IT) AND 2^(NS – IS)) AND (FR$(IS) = "R") THEN R$ = "$"
660 NEXT IS
670 NEXT J
680 IF IT = NT THEN R$ = "$"
690 SY$(IT) = R$ + T$(IT)
700 NEXT IT
710 PRINT
720 PRINT "Source Expectation of mean square"
730 PRINT" = = = = = = = = = = = = = = = = = = = = = = = = = = = = = = = =
    = ="
740 FOR IT = 1 TO NT – 1
750 PRINT T$(IT);TAB(9);
760 FOR IS = 1 TO NS
770 CV(IS) = 1
780 NEXT IS
790 FOR IS = 1 TO NS
800 IF (SUB(IT) AND 2^(NS – IS)) = 0 THEN 840
810 FOR JS = 1 TO NS
820 IF (SUB(IT) AND 2^(NS – IS)) = (SYM(IT) AND 2^(NS – JS)) THEN CV(IS) = 0
830 NEXT JS
840 NEXT IS
850 NP = 0
860 FOR JT = NT TO 1 STEP – 1
870 N = 1
880 FOR JS = 1 TO NS
890 IF CV(JS) <> 0 THEN N = N*TB(JT,JS)
900 NEXT JS
910 IF N = 0 THEN 970
920 IF (SYM(IT) AND SUB(JT)) <> SYM(IT) THEN 970
930 NP = NP + 1
940 IF NP > 1 THEN PRINT " +";
950 IF NP = 1 THEN PRINT SY$(JT);
960 IF NP > 1 THEN PRINT N;SY$(JT);
970 NEXT JT
980 PRINT
990 NEXT IT
1000 PRINT "E          $E"
1010 PRINT" = = = = = = = = = = = = = = = = = = = = = = = = = = = = = = = =
     = = = = = = = = = = = = = = = = = = = ="
1020 PRINT "NOTE: In expectations of mean squares,"
1030 PRINT "      # denotes average square of fixed effects, and "
1040 PRINT "      $ denotes variance component for random effects."
1050 PRINT
1060 END
1070 REM Accept and process model for Balanced Analysis
1080 INPUT "Enter model :"; M$
1090 REM Determine number of terms (NT)
1100 NT = 0
```

```
1110 FOR J = 1 TO LEN(M$)
1120 IF MID$(M$,J,1) = "(" THEN NT = NT + 1
1130 NEXT J
1140 REM Determine number of subscripts (NS)
1150 NS = 0:SW = 0
1160 FOR J = 1 TO LEN(M$)
1170 IF MID$(M$,J,1) = "E" OR MID$(M$,J,1) = "e" THEN SW = 1:GOTO 1200
1180 IF SW = 0 THEN 1200
1190 IF MID$(M$,J,1) <> "(" AND MID$(M$,J,1) <> ")" THEN NS = NS + 1
1200 NEXT J
1210 REM Determine effects and subscripts
1220 DIM E$(NS),S$(NS),SYM(NT),SUB(NT),T$(NT)
1230 SW = 0:NE = 0:MS = 0:ISYM = 1:ISUB = 1
1240 FOR J = 1 TO LEN(M$)
1250 C$ = MID$(M$,J,1)
1260 IF C$ = "(" THEN SW = 1
1270 IF C$ = "(" THEN ISYM = ISYM + 1
1280 IF C$ = ")" THEN SW = 0
1290 IF C$ = ")" THEN ISUB = ISUB + 1
1300 IF C$ = "(" OR C$ = ")" THEN 1510
1310 IF C$ = "e" OR C$ = "E" THEN 1510
1320 IF SW = 1 THEN 1430
1330 T$(ISYM) = T$(ISYM)+C$
1340 ES = 0
1350 FOR IE = 1 TO NE
1360 IF E$(IE) = C$ THEN ES = 1
1370 NEXT IE
1380 IF ES = 0 THEN NE = NE + 1:E$(NE) = C$
1390 FOR IE = 1 TO NE
1400 IF E$(IE) = C$ THEN SYM(ISYM) = SYM(ISYM) OR (2^(NS - IE))
1410 NEXT IE
1420 GOTO 1510
1430 SS = 0
1440 FOR IS = 1 TO MS
1450 IF S$(IS) = C$ THEN SS = 1
1460 NEXT IS
1470 IF SS = 0 THEN MS = MS + 1:S$(MS) = C$
1480 FOR IS = 1 TO MS
1490 IF S$(IS) = C$ THEN SUB(ISUB) = SUB(ISUB) OR (2^(NS - IS))
1500 NEXT IS
1510 NEXT J
1520 T$(NT) = "E":SYM(NT) = 0:IF NE < NS THEN E$(NS) = "E"
1530 RETURN
```

APPENDIX 8:

PROGRAM FOR CALCULATING GENOTYPIC AND PHENOTYPIC COVARIANCES

PROGRAM NAME: LINEAR.BAS

LANGUAGE: BASIC

PURPOSE:

This program calculates genotypic and phenotypic covariances as linear functions of estimated mean squares and cross-products. It also gives approximate standard errors of the genotypic and phenotypic covariances, and estimates of heritability and approximate standard errors of heritability for each trait.

DESCRIPTION:

This program reads mean squares and mean cross-products from an input file, prompts the user to indicate the contribution of each item to the estimate of genotypic and phenotypic variances and covariances, and then calculates the linear functions and their standard errors. Standard errors of linear functions of mean squares and mean cross-products as well as of estimates of heritability are calculated by the methods described in Chapter 4.

USER INSTRUCTIONS:

When this program is run, the user specifies the name of the file that contains the estimated mean squares and mean cross-products. This file must have a format such as that created by BACOVA1 (Appendix 5) or BACOVA2 (Appendix 6). For each source of variation, the program reads (1) the name of the term, (2) the degrees of freedom associated with the term, and (3) the $NV(NV + 1)/2$ mean squares and mean cross-products for that term.

The user will then be prompted to indicate the appropriate coefficients for calculating the genotypic and phenotypic covariances as linear functions of the mean cross-products. Thus, if the genotypic variance is given by (0.25 mean square $1 - 0.25$ mean square 2) and the phenotypic variance is estimated by (mean square 1), the user would specify the coefficients 0.25 and 1 for mean square 1 and the coefficients -0.25 and 0 for mean square 2. Coefficients required for various designs and various selection strategies are considered in Chapter 4.

The program will then calculate and print the estimated genotypic and phenotypic covariances and their approximate standard errors, as well as the heritability and its standard error for each trait. The estimated genotypic and phenotypic covariances are written to a file in a form suitable for subsequent input to HAYES (Appendix 9) and INDEX (Appendix 10).

PROGRAM LISTING:

```
10 REM Program to combine mean cross-products into estimates
20 REM of genotypic and phenotypic covariances
30 REM
40 REM Turn key definitions off, clear screen, prompt.
50 REM
60 KEY OFF:CLS
70 PRINT
80 PRINT" Linear functions of mean cross-products"
90 PRINT" = = = = = = = = = = = = = = = = = = = = = = = = = = = = = = = =
        = = = = = = = ="
```

```
100 PRINT "What is the name of the file"
110 INPUT "        that contains the mean cross-products "; FI$
120 PRINT
130 PRINT "What is the name of the file"
140 PRINT "        into which genotypic and phenotypic covariances"
150 INPUT "        are to be written "; FO$
160 PRINT
170 INPUT "How many variables were there in the analysis "; NV
180 PRINT
190 D = NV*(NV + 1)/2
200 DIM T$(40),DF(40),MCP(40,D),GC(40),PC(40),G(D),P(D),GE(D),PE(D)
210 DIM GPE(D),H2(D),H2E(D)
220 REM
230 REM Read source name, d.f., and mean cross-products.
240 REM
250 OPEN FI$ FOR INPUT AS #1
260 NT = 0
270 IF EOF(1) THEN 420
280 NT = NT + 1
290 INPUT #1,T$(NT)
300 INPUT #1,DF(NT)
310 JK = 0
320 FOR J = 1 TO NV
330 FOR K = 1 TO J
340 JK = JK + 1
350 INPUT #1,MCP(NT,JK)
360 NEXT K
370 NEXT J
380 GOTO 270
390 REM
400 REM Print mean cross-products on screen.
410 REM
420 CLS
430 PRINT "Mean cross products read from "; FI$
440 PRINT:PRINT"Source df"
450 FOR IT = 1 TO NT
460 PRINT T$(IT); TAB(8); DF(IT)
470 NEXT IT
480 REM
490 REM Prompt for coefficients that indicate contribution
500 REM of each cross-product to genotypic and phenotypic
510 REM components of covariance.
520 REM
530 PRINT
540 PRINT"You must specify two coefficients for each source"
550 PRINT"The first indicates contribution to genotypic covariances,"
560 PRINT"the second to phenotypic covariances."
570 PRINT
580 FOR IT = 1 TO NT
590 PRINT "G coefficient for "; T$(IT);
600 INPUT GC(IT)
```

```
610 PRINT "P coefficient for "; T$(IT);
620 INPUT PC(IT)
630 PRINT
640 NEXT IT
650 CLS
660 REM
670 REM Check that coefficients are entered correctly.
680 REM
690 PRINT "Source        df      Genotypic      Phenotypic"
700 FOR IT = 1 TO NT
710 PRINT T$(IT); TAB(11); DF(IT); TAB(19); GC(IT); TAB(32); PC(IT)
720 NEXT IT
730 PRINT:INPUT" OK?     (1 = yes, 0 = no)     ";SW
740 IF SW = 1 THEN 820
750 PRINT "Which item is in error (1 to "; NT;") ";
760 INPUT IT
770 PRINT "G coefficient for "; T$(IT);
780 INPUT GC(IT)
790 PRINT "P coefficient for "; T$(IT);
800 INPUT PC(IT)
810 GOTO 650
820 CLOSE 1
830 REM
840 REM Calculate, save and print linear functions of
850 REM mean squares and mean cross-products, as well
860 REM as heritabilities and corresponding standard
870 REM errors.
880 REM
890 JK = 0
900 FOR J = 1 TO NV
910 FOR K = 1 TO J
920 JK = JK + 1
930 FOR IT = 1 TO NT
940 G(JK) = G(JK) + MCP(IT,JK)*GC(IT)
950 P(JK) = P(JK) + MCP(IT,JK)*PC(IT)
960 NEXT IT
970 NEXT K
980 NEXT J
990 OPEN FO$ FOR OUTPUT AS #2
1000 JK = 0
1010 FOR J = 1 TO NV
1020 FOR K = 1 TO J
1030 JK = JK + 1
1040 PRINT #2,G(JK),P(JK)
1050 NEXT K
1060 NEXT J
1070 PRINT
1080 PRINT"      I      J      G(I,J)      SE of G(I,J)      P(I,J)      SE of
     P(I,J)      H2      SE of H2"
1090 PRINT" = = = = = = = = = = = = = = = = = = = = = = = = = = = = = = =
     = = = = = = = = = = = = = = = = = = = = = = = = = = = = = = = = =
     = = = = = = = = = = = = ="
```

```
1100 PRINT
1110 JK = 0
1120 FOR J = 1 TO NV
1130 FOR K = 1 TO J
1140 JK = JK + 1
1150 JJ = 0
1160 FOR J1 = 1 TO J
1170 FOR J2 = 1 TO J1
1180 JJ = JJ + 1
1190 NEXT J2
1200 NEXT J1
1210 KK = 0
1220 FOR K1 = 1 TO K
1230 FOR K2 = 1 TO K1
1240 KK = KK + 1
1250 NEXT K2
1260 NEXT K1
1270 FOR IT = 1 TO NT
1280 GE(JK) = GE(JK) +
     GC(IT)*GC(IT)*(MCP(IT,JJ)*MCP(IT,KK) + MCP(IT,JK)*MCP(IT,JK))/
     (DF(IT) + 2)
1290 PE(JK) = PE(JK) +
     PC(IT)*PC(IT)*(MCP(IT,JJ)*MCP(IT,KK) + MCP(IT,JK)*MCP(IT,JK))/(DF(IT) + 2)
1300 GPE(JK) = GPE(JK) +
     GC(IT)*PC(IT)*(MCP(IT,JJ)*MCP(IT,KK) + MCP(IT,JK)*MCP(IT,JK))/(DF(IT) + 2)
1310 NEXT IT
1320 GE(JK) = SQR(GE(JK))
1330 PE(JK) = SQR(PE(JK))
1340 IF J <> K THEN 1390
1350 H2(JK) = G(JK)/P(JK)
1360 H2E(JK) = GE(JK)*GE(JK) −
     2!*H2(JK)*GPE(JK) + H2(JK)*H2(JK)*PE(JK)*PE(JK)
1370 H2E(JK) = SQR(H2E(JK)/P(JK)/P(JK))*100
1380 H2(JK) = H2(JK)*100
1390 PRINT  USING"  ##     ##     ######.###     #####.###     ###
     ###.###     #####.##     ###.#     ###.#";
     J,K,G(JK),GE(JK),P(JK),PE(JK),H2(JK),H2E(JK)
1400 NEXT K
1410 NEXT J
1420 PRINT
1430 CLOSE 2
1440 END
```

APPENDIX 9:

PROGRAM FOR CALCULATING HAYES-HILL TRANSFORMATION OF GENOTYPIC AND PHENOTYPIC COVARIANCE MATRICES

PROGRAM NAME: HAYES.BAS

LANGUAGE: BASIC

PURPOSE:

This program is designed to perform the transformation recommended by Hayes and Hill in order to show the sampling properties of the estimated genotypic and phenotypic covariance matrices.

DESCRIPTION:

Estimated genotypic and phenotypic covariance matrices are transformed so that the genotypic and phenotypic correlations on the transformed scale are zero, and phenotypic variances on the transformed scale are 1.0 for all variables. If estimated genotypic variances on the transformed scale (printed as eigenvalues) are less than zero or greater than one, there is an indication that the sampling has resulted in estimates of heritability that are outside the range of 0 to 1 or estimates of partial correlations outside the acceptable range of -1 to $+1$. In any event, the estimated covariance matrices must be viewed with caution and their use as a basis for developing a selection index must be questioned.

USER INSTRUCTIONS:

The program first prompts for the name of the file that contains the genotypic and phenotypic covariances stored in the form produced by LINEAR (Appendix 8) or by the simulation program MNORM (Appendix 3). The input file must therefore contain $G(1,1),P(1,1),G(2,1),P(2,1,)G(2,2),P(2,2),G(3,1),P(3,1),G(3,2),P(3,2)$, etc., without any identification. To facilitate proper input, the user will then be prompted to indicate how many variables are included in the genotypic and phenotypic covariance matrices.

The program calculates the necessary transformation and prints the eigenvalues and their corresponding eigenvectors. Acceptable eigenvalues are between 0 and 1. Eigenvectors are printed in case the user wishes to study the form of the transformation. They are not part of the test of the adequacy of the estimated covariances.

PROGRAM LISTING:

```
10 REM Program for applying transformation to genotypic
20 REM and phenotypic covariance matrices as recommended
30 REM by Hayes, J. F., and Hill, W. G., A reparamterization
40 REM of a genetic selection index to locate its sampling
50 REM properties, Biometrics,36,237,1980.
60 REM
70 REM Turn key definitions off, clear screen, prompt.
80 REM
90 KEY OFF:CLS
100 PRINT
110 PRINT "Program for calculating Hayes and Hill"
120 PRINT "transformation of covariance matrices"
130 PRINT" = = = = = = = = = = = = = = = = = = = = = = = = = = = = = = = =
    = = = = = = ="
```

```
140 PRINT
150 REM
160 REM Read genotypic and phenotypic covariances from file.
170 REM G and P matrices stored in lower triangular form,
180 REM i.e., G(1,1),P(1,1),G(2,1),P(2,1),G(2,2),P(2,2),
190 REM G(3,1),P(3,1),G(3,2),P(3,2),G(3,3),P(3,3),etc.
200 REM
210 PRINT "Enter name of file that contains genotypic"
220 INPUT "         and phenotypic covariances "; INFIL$
230 PRINT
240 OPEN INFIL$ FOR INPUT AS #1
250 INPUT "How many variables "; N
260 PRINT
270 DIM AMAT(N,N),BMAT(N,N),TORTH(N,N),DL(N),DIAG(N)
280 PRINT
290 PRINT"                                                          Covariance"
300 PRINT"Variable 1 Variable 2 Genotypic Phenotypic"
310 PRINT" = = = = = = = = = = = = = = = = = = = = = = = = = = = = = = = =
    = = = = = = = = = = ="
320 FOR I = 1 TO N
330 FOR J = 1 TO N
340 IF J > I THEN 370
350 INPUT #1,AMAT(I,J),BMAT(I,J)
360 PRINT USING"    ####        ####       #####.###     #####.##
    #"; I,J,AMAT(I,J),BMAT(I,J)
370 NEXT J
380 NEXT I
390 FOR I = 1 TO N
400 FOR J = 1 TO N
410 IF J > I THEN AMAT(I,J) = AMAT(J,I)
420 IF J > I THEN BMAT(I,J) = BMAT(J,I)
430 NEXT J
440 NEXT I
450 REM
460 REM Call subroutine to convert to a symmetric
470 REM eigenproblem.
480 REM
490 GOSUB 2530
500 REM
510 REM Call subroutine to calculate eigenvalues
520 REM and eigenvectors of a matrix.
530 REM
540 GOSUB 840
550 REM
560 REM Call subroutine to convert eigenvalues
570 REM back to appropriate form.
580 REM
590 GOSUB 3020
600 REM
610 REM Print results of the transformation.
620 REM
```

```
630 PRINT
640 PRINT" Eigenvalues are :"
650 FOR I = 1 TO N
660 PRINT USING " ##        ###.#####"; I,DIAG(I)
670 NEXT I
680 PRINT:PRINT "Corresponding eigenvectors are :"
690 FOR I = 1 TO N
700 FOR J = 1 TO N
710 PRINT USING " ###.###";TORTH(I,J);
720 NEXT J
730 PRINT
740 NEXT I
750 PRINT
760 CLOSE #1
770 END
780 REM
790 REM Subroutine tred 2 for eigenvalues and eigenvectors.
800 REM Martin, R. S., Reinsch, C., and Wilkinson, J. H.,
810 REM Householder's tridiagonolization of a
820 REM symmetric matrix, Numer. Math.,11,181,1968.
830 REM
840 FOR I = 1 TO N
850 FOR J = 1 TO N
860 TORTH(I,J) = AMAT(I,J)
870 NEXT J
880 NEXT I
890 IF N = 1 THEN 1370
900 FOR I = N TO 2 STEP −1
910 L = I − 1
920 H = 0
930 SCALE = 0
940 IF L < 2 THEN 990
950 FOR K = 1 TO L
960 SCALE = SCALE + ABS(TORTH(I,K))
970 NEXT K
980 IF SCALE > 0 THEN 1010
990 SUB(I) = TORTH(I,L)
1000 GOTO 1350
1010 FOR K = 1 TO L
1020 TORTH(I,K) = TORTH(I,K)/SCALE
1030 H = H + TORTH(I,K) * TORTH(I,K)
1040 NEXT K
1050 F = TORTH(I,L)
1060 G = SQR(H)
1070 IF F > 0 THEN G = −G
1080 SUB(I) = SCALE * G
1090 H = H − F * G
1100 TORTH(I,L) = F − G
1110 F = 0
1120 FOR J = 1 TO L
1130 TORTH(J,I) = TORTH(I,J)/H
```

```
1140 G = 0
1150 FOR K = 1 TO J
1160 G = G + TORTH(J,K) * TORTH(I,K)
1170 NEXT K
1180 JP1 = J + 1
1190 IF L < JP1 THEN 1230
1200 FOR K = JP1 TO L
1210 G = G + TORTH(K,J) * TORTH(I,K)
1220 NEXT K
1230 SUB(J) = G/H
1240 F = F + SUB(J) * TORTH(I,J)
1250 NEXT J
1260 HH = F/(H + H)
1270 FOR J = 1 TO L
1280 F = TORTH(I,J)
1290 G = SUB(J) − HH * F
1300 SUB(J) = G
1310 FOR K = 1 TO J
1320 TORTH(J,K) = TORTH(J,K) − F * SUB(K) − G * TORTH(I,K)
1330 NEXT K
1340 NEXT J
1350 DIAG(I) = H
1360 NEXT I
1370 DIAG(1) = 0
1380 SUB(1) = 0
1390 FOR I = 1 TO N
1400 L = I − 1
1410 IF DIAG(I) = 0 THEN 1510
1420 FOR J = 1 TO L
1430 G = 0
1440 FOR K = 1 TO L
1450 G = G + TORTH(I,K) * TORTH(K,J)
1460 NEXT K
1470 FOR K = 1 TO L
1480 TORTH(K,J) = TORTH(K,J) − G * TORTH(K,I)
1490 NEXT K
1500 NEXT J
1510 DIAG(I) = TORTH(I,I)
1520 TORTH(I,I) = 1
1530 IF L < 1 THEN 1580
1540 FOR J = 1 TO L
1550 TORTH(I,J) = 0
1560 TORTH(J,I) = 0
1570 NEXT J
1580 NEXT I
1590 REM
1600 REM Subroutine tql 2.
1610 REM Bowdler, H., Martin, R. S., Reinsch, C., and
1620 REM Wilkinson, J. H., The QR and QL algorithms
1630 REM for symmetric matrices, Numer. Math.,11,293,1968.
1640 REM
```

```
1650 REM Variable MACHEP is a machine dependent variable
1660 REM and may have to be modified for other machines.
1670 REM MACHEP is the smallest number for which
1680 REM 1 + MACHEP > 1.
1690 REM
1700 MACHEP = .0000001
1710 IERR = 0
1720 IF N = 1 THEN 2520
1730 FOR I = 2 TO N
1740 SUB(I − 1) = SUB(I)
1750 NEXT I
1760 F = 0
1770 B = 0
1780 SUB(N) = 0
1790 FOR L = 1 TO N
1800 J = 0
1810 H = MACHEP * (ABS(DIAG(L)) + ABS(SUB(L)))
1820 IF B < H THEN B = H
1830 FOR M = L TO N
1840 IF M = N THEN 1870
1850 IF ABS(SUB(M)) < B THEN 1870
1860 NEXT M
1870 IF M = L THEN 2300
1880 IF J = 30 THEN 2510
1890 J = J + 1
1900 L1 = L + 1
1910 G = DIAG(L)
1920 P = (DIAG(L1) − G)/(2 * SUB(L))
1930 R = SQR(P * P + 1) : SR = R : IF P < 0 THEN SR = −SR
1940 DIAG(L) = SUB(L)/(P + SR)
1950 H = G − DIAG(L)
1960 FOR I = L1 TO N
1970 DIAG(I) = DIAG(I) − H
1980 NEXT I
1990 F = F + H
2000 P = DIAG(M)
2010 C = 1
2020 S = 0
2030 MML = M − L
2040 FOR I = M − 1 TO L STEP −1
2050 G = C * SUB(I)
2060 H = C * P
2070 IF ABS(P) < ABS(SUB(I)) THEN 2140
2080 C = SUB(I) / P
2090 R = SQR(C * C + 1)
2100 SUB(I + 1) = S * P * R
2110 S = C/R
2120 C = 1/R
2130 GOTO 2190
2140 C = P/SUB(I)
2150 R = SQR(C * C + 1)
```

```
2160 SUB(I+1) = S * SUB(I) * R
2170 S = 1 / R
2180 C = C * S
2190 P = C * DIAG(I) − S * G
2200 DIAG(I+1) = H + S * (C * G + S * DIAG(I))
2210 FOR K = 1 TO N
2220 H = TORTH(K,I+1)
2230 TORTH(K,I+1) = S * TORTH(K,I) + C * H
2240 TORTH(K,I) = C * TORTH(K,I) − S * H
2250 NEXT K
2260 NEXT I
2270 SUB(L) = S * P
2280 DIAG(L) = C * P
2290 IF ABS(SUB(L)) > B THEN 1880
2300 DIAG(L) = DIAG(L) + F
2310 NEXT L
2320 FOR II = 2 TO N
2330 I = II − 1
2340 K = I
2350 P = DIAG(I)
2360 FOR J = II TO N
2370 IF DIAG(J) >= P THEN 2400
2380 K = J
2390 P = DIAG(J)
2400 NEXT J
2410 IF K = I THEN 2490
2420 DIAG(K) = DIAG(I)
2430 DIAG(I) = P
2440 FOR J = 1 TO N
2450 P = TORTH(J,I)
2460 TORTH(J,I) = TORTH(J,K)
2470 TORTH(J,K) = P
2480 NEXT J
2490 NEXT II
2500 GOTO 2520
2510 IERR = L
2520 RETURN
2530 REM Subroutine to convert Ax = lambda Bx to
2540 REM symmetric eigenproblem.
2550 REM Martin, R. S., and Wilkinson, J. H., Reduction
2560 REM of the symmetric eigenproblem Ax = 1Bx and related
2570 REM problems to standard form. Numer. Math.,11,110,1968.
2580 REM
2590 REM reduc − 1
2600 REM
2610 FOR I = 1 TO N
2620 FOR J = I TO N
2630 X = BMAT(I,J)
2640 IF I <= 1 THEN 2680
2650 FOR K = I−1 TO 1 STEP −1
2660 X = X − BMAT(I,K)*BMAT(J,K)
```

```
2670 NEXT K
2680 IF I = J AND X < = 0 THEN PRINT "B is non-positive definite":STOP
2690 IF I = J THEN Y = SQR(X):DL(I) = Y
2700 IF I <> J THEN BMAT(J,I) = X/Y
2710 NEXT J
2720 NEXT I
2730 FOR I = 1 TO N
2740 Y = DL(I)
2750 FOR J = I TO N
2760 X = AMAT(I,J)
2770 IF I < = 1 THEN 2810
2780 FOR K = I-1 TO 1 STEP -1
2790 X = X - BMAT(I,K) * AMAT(J,K)
2800 NEXT K
2810 AMAT(J,I) = X/Y
2820 NEXT J
2830 NEXT I
2840 FOR J = 1 TO N
2850 FOR I = J TO N
2860 X = AMAT(I,J)
2870 IF I < = J THEN 2910
2880 FOR K = I-1 TO J STEP -1
2890 X = X - AMAT(K,J) * BMAT(I,K)
2900 NEXT K
2910 IF J < = 1 THEN 2950
2920 FOR K = J-1 TO 1 STEP -1
2930 X = X - AMAT(J,K) * BMAT(I,K)
2940 NEXT K
2950 AMAT(I,J) = X/DL(I)
2960 NEXT I
2970 NEXT J
2980 RETURN
2990 REM
3000 REM Subroutine to convert eigenvectors back to
3010 REM correct form.
3020 REM
3030 REM rebak-a      (See reference for reduc - 1 above)
3040 REM
3050 FOR J = 1 TO N
3060 FOR I = N TO 1 STEP -1
3070 X = TORTH(I,J)
3080 IF I = N THEN 3120
3090 FOR K = I+1 TO N
3100 X = X - BMAT(K,I) * TORTH(K,J)
3110 NEXT K
3120 TORTH(I,J) = X/DL(I)
3130 NEXT I
3140 NEXT J
3150 RETURN
```

APPENDIX 10:

PROGRAM FOR ESTIMATING COEFFICIENTS FOR AN OPTIMUM SELECTION INDEX

PROGRAM NAME: INDEX.BAS

LANGUAGE: BASIC

PURPOSE:

This program is designed to estimate the index coefficients for the optimum index procedure of Smith.

DESCRIPTION:

Estimates of phenotypic covariances are stored in the matrix **P** and estimates of genotypic covariances are stored in the matrix **G**. The user specifies the relative economic value of each trait and these are stored in the vector **a**. The vector of index coefficients, **b**, are those given by solving the system of equations $\mathbf{Pb} = \mathbf{Ga}$ to get $\mathbf{b} = \mathbf{P}^{-1}\mathbf{Ga}$. The equations are solved by a modified Gauss elimination procedure.

In addition to calculating the estimated index coefficients, the program calculates and prints the response in overall worth, and in each trait, that would be expected if the best 35% were selected on the basis of the index scores (standardized selection differential = 1.0).

USER INSTRUCTIONS:

The user is first prompted to enter the number of variables and the name of the file that contains the estimates of genotypic and phenotypic covariances. The file must contain the genotypic and phenotypic covariances in the form of lower triangular matrices without identification (see Appendix 3 User Instructions).

The program will then prompt the user to enter the relative economic value for each trait. If index selection is for improving a single trait, the relative economic value will be entered as 1.0 for that trait and 0.0 for all other traits. The program prints the equations that require solution as well as the index coefficients that are the solutions to those equations.

Expected response to selection is calculated and printed for overall genotypic worth and for the genotypic values of each trait. The user can study these estimates to get an idea of whether or not the limitations imposed by the structure of the population from which the variances and covariances were estimated will seriously reduce the anticipated responses. The user may wish to run this program again with some other set of relative economic values to assess their impact on expected responses.

PROGRAM LISTING:

```
10 REM Program to calculate index coefficients b = P(-1)G
20 REM and expected responses r = Cov(g,I)/sd(I), for
30 REM an estimated optimum index.
40 REM
50 REM Turn off key definitions and clear screen.
60 REM
70 KEY OFF:CLS
80 REM
90 REM Read genotypic and phenotypic covariance matrices
100 REM Covariances stored in lower triangular mode,
```

```
110 REM i.e., G(1,1),P(1,1),G(2,1),P(2,1),G(2,2),P(2,2)
120 REM G(3,1),P(3,1),G(3,2),P(3,2),G(3,3),P(3,3),etc.
130 REM
140 PRINT
150 PRINT "Program for calculating optimum selection index"
160 PRINT "= = = = = = = = = = = = = = = = = = = = = = = = = = = = = = =
    = = = = = = = = = = = = = = = = ="
170 PRINT
180 INPUT "How many variables "; NV
190 PRINT
200 PRINT "What is name of file that contains"
210 INPUT "        the covariance matrices "; FI$
220 OPEN FI$                                         FOR INPUT AS #1
230 DIM G(NV,NV),P(NV,NV),PI(NV,NV),A(NV),B(NV),RH(NV)
240 JK = 0
250 FOR J = 1 TO NV
260 FOR K = 1 TO J
270 INPUT #1,G(J,K),P(J,K)
280 IF J <> K THEN G(K,J) = G(J,K):P(K,J) = P(J,K)
290 NEXT K
300 NEXT J
310 REM
320 REM Prompt for relative economic values
330 REM
340 PRINT
350 FOR IV = 1 TO NV
360 PRINT "What is relative economic value for trait "; IV;" ";
370 INPUT A(IV)
380 NEXT IV
390 REM
400 REM Calculate right hand side of equations
410 REM
420 FOR IV = 1 TO NV
430 RH(IV) = 0
440 FOR JV = 1 TO NV
450 RH(IV) = RH(IV) + A(JV) * G(IV,JV)
460 NEXT JV
470 NEXT IV
480 REM
490 REM Solve equations by Gauss-Jordan elimination
500 REM Method is modified from that given by
510 REM Miller, A. R., BASIC programs for scientists
520 REM and engineers, SYBEX Inc.,Berkeley, CA, 1981.
530 REM
540 DIM W%(NV,3),AA(NV,NV),BB(NV,NV)
550 FOR IV = 1 TO NV
560 FOR JV = 1 TO NV
570 AA(IV,JV) = P(IV,JV)
580 BB(IV,JV) = P(IV,JV)
590 NEXT JV
600 B(IV) = RH(IV)
```

```
610 W%(IV,3) = 0
620 NEXT IV
630 FOR IV = 1 TO NV
640 DT = 1
650 BG = 0
660 FOR JV = 1 TO NV
670 IF W%(JV,3) = 1 THEN 750
680 FOR KV = 1 TO NV
690 IF W%(KV,3) > 1 THEN PRINT "Matrix cannot be inverted ":STOP
700 IF W%(KV,3) = 1 THEN 740
710 IR = JV
720 IC = KV
730 BG = ABS(BB(JV,KV))
740 NEXT KV
750 NEXT JV
760 W%(IC,3) = W%(IC,3) + 1
770 W%(IV,1) = IR
780 W%(IV,2) = IC
790 IF IR = IC THEN 890
800 DT = - DT
810 FOR LV = 1 TO NV
820 H = BB(IR,LV)
830 BB(IR,LV) = BB(IC,LV)
840 BB(IC,LV) = H
850 NEXT LV
860 H = B(IR)
870 B(IR) = B(IC)
880 B(IC) = H
890 PV = BB(IC,IC)
900 DT = DT * PV
910 BB(IC,IC) = 1
920 FOR LV = 1 TO NV
930 BB(IC,LV) = BB(IC,LV)/PV
940 NEXT LV
950 B(IC) = B(IC)/PV
960 FOR MV = 1 TO NV
970 IF MV = IC THEN 1040
980 H = BB(MV,IC)
990 BB(MV,IC) = 0
1000 FOR LV = 1 TO NV
1010 BB(MV,LV) = BB(MV,LV) - BB(IC,LV)*H
1020 NEXT LV
1030 B(MV) = B(MV) - B(IC)*H
1040 NEXT MV
1050 NEXT IV
1060 FOR IV = 1 TO NV
1070 LV = NV - IV + 1
1080 IF W%(LV,1) = W%(LV,2) THEN 1160
1090 IR = W%(LV,1)
1100 IC = W%(LV,2)
1110 FOR KV = 1 TO NV
```

```
1120 H = BB(KV,IR)
1130 BB(KV,IR) = BB(KV,IC)
1140 BB(KV,IC) = H
1150 NEXT KV
1160 NEXT IV
1170 FOR KV = 1 TO NV
1180 IF W%(KV,3) <> 1 THEN PRINT "Equations cannot be solved ":STOP
1190 NEXT KV
1200 REM
1210 REM Print equations
1220 REM
1230 PRINT
1240 PRINT"Equations were:"
1250 FOR IV = 1 TO NV
1260 FOR JV = 1 TO NV
1270 PRINT USING " ######.### "; P(IV,JV);
1280 NEXT JV
1290 PRINT USING " = ######.###"; RH(IV)
1300 NEXT IV
1310 PRINT
1320 PRINT USING "Determinant = #####.###"; DT
1330 PRINT
1340 PRINT "Inverse phenotypic covariance matrix"
1350 FOR IV = 1 TO NV
1360 FOR JV = 1 TO NV
1370 PRINT USING " ###.###### "; BB(IV,JV);
1380 NEXT JV
1390 PRINT
1400 NEXT IV
1410 REM
1420 REM Pause before printing remainder
1430 REM
1440 PRINT
1450 INPUT "Press return to continue "; C$
1460 CLS
1470 PRINT
1480 PRINT "Index coefficients:"
1490 FOR IV = 1 TO NV
1500 PRINT USING " b(##) = ####.### "; IV,B(IV)
1510 NEXT IV
1520 PRINT
1530 REM
1540 REM Calculate expected responses
1550 REM
1560 DIM R(NV)
1570 SI = 0
1580 WI = 0
1590 FOR IV = 1 TO NV
1600 R(IV) = 0
1610 FOR JV = 1 TO NV
1620 R(IV) = R(IV) + B(JV) * G(IV,JV)
```

```
1630 SI = SI + B(IV) * B(JV) * P(IV,JV)
1640 WI = WI + A(IV) * B(JV) * G(IV,JV)
1650 NEXT JV
1660 NEXT IV
1670 SI = SQR(SI)
1680 RI = WI/SI
1690 FOR IV = 1 TO NV
1700 R(IV) = R(IV)/SI
1710 NEXT IV
1720 PRINT "Expected response for selection differential of"
1730 PRINT "one standard deviation (based on index scores):"
1740 PRINT
1750 PRINT USING "Response for overall genotypic worth = ####.###"; RI
1760 PRINT
1770 FOR IV = 1 TO NV
1780 PRINT USING "Response for trait number ## = ####.###"; IV; R(IV)
1790 NEXT IV
1800 PRINT
1810 END
```

APPENDIX 11:

PROGRAM FOR CALCULATION OF, AND RANKING ACCORDING TO, INDEX SCORES

PROGRAM NAME: SCORES.BAS

LANGUAGE: BASIC

PURPOSE:

This program can be used to calculate index scores for a number of genotypes given the phenotypic values for a number of traits for each genotype and the weights (index coefficients) to be attached to each trait.

DESCRIPTION:

The index score for each genotype is calculated by multiplying the phenotypic value for each trait by its corresponding index coefficient, and summing over traits for each genotype. Genotypes are then ordered from those having the largest scores to those having the smallest scores. Identification numbers, data for each trait, and index scores are printed for each genotype.

USER INSTRUCTIONS:

When the program is run, the user will be prompted to indicate the name of the file that contains the average phenotypic values for each genotype, along with an identification number for each genotype. The user must also specify the number of traits measured on each genotype and the number of genotypes included in the input file.

Once the program has read the data from the input file, the user will be prompted to give the index coefficients for each trait. The program then calculates the index score for each genotype, sorts genotypes from largest index score to smallest, and prints the identification, data, and index score for each genotype on a printer.

Index coefficients used for this program may be those calculated using INDEX (Appendix 10) or developed by some other method. For example, the user may choose to weight each trait according to its relative economic value, its heritability, or the product of the two as discussed in Chapter 8.

PROGRAM LISTING:

```
 10   REM Program to calculate linear index scores.
 20   REM This program reads an identification number
 30   REM for each record (genotype) and one value for
 40   REM each trait. It then prompts for the weights
 50   REM (index coefficients) to be applied to each
 60   REM trait, calculates the index scores, sorts
 70   REM the records from high to low index score,
 80   REM and prints the sorted records.
 90   REM
100   REM Turn key definitions off and clear screen.
110   REM
120   KEY OFF:CLS
130   REM
140   REM Prompt for file name, number of variables, and
150   REM number of genotypes.
```

```
160  REM
170  PRINT
180  PRINT" Program to calculate linear index scores "
190  PRINT
200  INPUT "What is the name of the file containing the data "; FI$
210  PRINT
220  INPUT "How many variables "; NV
230  INPUT "How many genotypes "; NR
240  PRINT
250  REM
260  REM Read data from file
270  REM
280  OPEN FI$ FOR INPUT AS #1
290  DIM ID(NR),X(NR,NV),SC(NR),B(NV)
300  FOR IR = 1 TO NR
310  INPUT #1,ID(IR)
320  FOR IV = 1 TO NV
330  INPUT #1,X(IR,IV)
340  NEXT IV
350  NEXT IR
360  REM
370  REM Prompt for index coefficients.
380  REM
390  FOR IV = 1 TO NV
400  PRINT USING "What is coefficient for variable ## "; IV;
410  INPUT B(IV)
420  NEXT IV
430  PRINT
440  REM
450  REM Calculate index scores.
460  REM
470  FOR IR = 1 TO NR
480  SC(IR) = 0
490  FOR IV = 1 TO NV
500  SC(IR) = SC(IR) + X(IR,IV)*B(IV)
510  NEXT IV
520  NEXT IR
530  REM
540  REM Sort from high to low index score
550  REM Use simple bubble sort routine.
560  REM
570  FOR IR = 1 TO NR - 1
580  FOR JR = IR + 1 TO NR
590  IF SC(JR) < = SC(IR) THEN 710
600  T = ID(IR)
610  ID(IR) = ID(JR)
620  ID(JR) = T
630  FOR IV = 1 TO NV
640  T = X(IR,IV)
650  X(IR,IV) = X(JR,IV)
660  X(JR,IV) = T
```

```
670   NEXT IV
680   T = SC(IR)
690   SC(IR) = SC(JR)
700   SC(JR) = T
710   NEXT JR
720   NEXT IR
730   REM
740   REM Print index coefficients and data on printer.
750   REM
760   PRINT "List data to printer in sorted order"
770   LPRINT" Data listing with index scores based on the following"
780   LPRINT" index coefficients"
790   LPRINT
800   LPRINT" Trait number Index coefficient"
810   FOR IV = 1 TO NV
820   LPRINT USING "        ##        ####.###"; IV,B(IV)
830   NEXT IV
840   LPRINT
850   LPRINT "Identif.      Data";:LPRINT TAB(55),"Score"
860   LPRINT
870   FOR IR = 1 TO NR
880   LPRINT USING " ####      "; ID(IR);
890   FOR IV = 1 TO NV
900   LPRINT USING "     ####.###"; X(IR,IV);
910   IF INT(IV/5) * 5 = IV THEN LPRINT:LPRINT TAB(10)
920   NEXT IV
930   LPRINT TAB(60);
940   LPRINT USING "     ####.###"; SC(IR)
950   NEXT IR
960   CLOSE 1
970   LPRINT "= = = = = = = = = = = = = = = = = = = = = = = = = = = =
      = = = = = = = = = = = = = = = = = = = = = = = = = = = ="
980   END
```

APPENDIX 12:

PROGRAM FOR WEIGHT-FREE RANKING OF GENOTYPES

PROGRAM NAME: ELSTON.BAS

LANGUAGE: BASIC

PURPOSE:
  This program is designed to calculate Elston's multiplicative index for each genotype and to print the data and a standardized score in ranked order.

DESCRIPTION:
  If smaller values of any trait are preferred, all values for that trait are multiplied by -1 so that large values on the transformed scale are best. Data for each trait are then adjusted by subtracting $k = (n*\text{minimum} - \text{maximum})/(n - 1)$ where n is the number of genotypes to be ranked. This has the effect of translating all traits to a similar location while preventing any trait from having an adjusted value of 0. The score for each genotype is then calculated as the product of the adjusted values for each trait. Genotypes are then ranked from highest to lowest score. Identification number, data for each trait, and the standardized score for each genotype are printed on a printer.

USER INSTRUCTIONS:
  This program is similar to SCORES (Appendix 11) in that the user is prompted for the name of the file that contains the identification numbers and average phenotypic values for each genotype, as well as for the number of traits measured on each genotype and the number of genotypes included in the file. The program will then calculate minimum, maximum, and average values for each trait and display them while prompting the user to indicate whether high or low values are preferred for that trait.
  The program will then carry out the calculations required for Elston's multiplicative index and print the identification, data, and standardized (0 to 100) scores for the sorted genotypes. No attempt is made to transform the data, other than to subtract a minimum value so that all variables have similar location. It is presumed that the distributions of all variables are similar, at least in the number of modes. If desired, data for particular traits could be transformed as they are read from the input file. For example, if one wished to transform the third trait to a logarithmic scale, it would be necessary to insert the statement

$$395 \text{ IF } IV = 3 \text{ THEN } X(IR,IV) = LOG(X(IR,IV))$$

into the program before it is run.

PROGRAM LISTING:
```
  10 REM Program for calculating weight-free index of
  20 REM Elston, R. C., A weight-free index for the purpose
  30 REM of ranking or selection with respect to several
  40 REM traits at a time, Biometrics,29,85,1963.
  50 REM
  60 REM Program reads identification number for each
  70 REM record (genotype) and data for all variables
  80 REM from a data file. User is prompted to indicate
  90 REM whether high or low values are preferred for
```

```
100 REM each variable. Elston's index scores are
110 REM calculated, genotypes are sorted from high
120 REM to low, and results are printed. This
130 REM program does not transform data to make
140 REM distributions more similar.
150 REM
160 REM Turn key definition off and clear screen
170 REM
180 KEY OFF:CLS
190 PRINT
200 PRINT " Weight-free index (Elston) for ranking genotypes"
210 PRINT "= = = = = = = = = = = = = = = = = = = = = = = = = = = = = = =
    = = = = = = = = = = = = = = = = = ="
220 PRINT
230 PRINT "What is the name of the file containing the data ";
240 INPUT FI$
250 PRINT
260 INPUT "How many variables "; NV
270 PRINT
280 INPUT "How many genotypes "; NR
290 PRINT
300 DIM ID(NR),X(NR,NV),Y(NR,NV),SC(NR),HI(NV),LO(NV),AV(NV)
310 REM
320 REM Read data from file
330 REM
340 OPEN FI$ FOR INPUT AS #1
350 FOR IR = 1 TO NR
360 SC(IR) = 1
370 INPUT #1,ID(IR)
380 FOR IV = 1 TO NV
390 INPUT #1,X(IR,IV)
400 Y(IR,IV) = X(IR,IV)
410 NEXT IV
420 NEXT IR
430 REM
440 REM Calculate minimum, maximum and average for each variable.
450 REM
460 FOR IV = 1 TO NV
470 HI(IV) = X(1,IV)
480 LO(IV) = X(1,IV)
490 AV(IV) = 0!
500 FOR IR = 1 TO NR
510 IF X(IR,IV) < LO(IV) THEN LO(IV) = X(IR,IV)
520 IF X(IR,IV) > HI(IV) THEN HI(IV) = X(IR,IV)
530 AV(IV) = AV(IV) + X(IR,IV)/NR
540 NEXT IR
550 REM
560 REM Determine whether high or low values preferred.
570 REM If low, change sign.
580 REM
590 PRINT
```

```
600 PRINT USING "For variable ##, the minimum is ####.###"; IV,LO(IV)
610 PRINT USING "          the maximum is ####.###"; HI(IV)
620 PRINT USING "          and the average is ####.###"; AV(IV)
630 PRINT
640 INPUT "Are high (1) or low (0) values preferred for this trait";SW
650 IF SW = 1 THEN 730
660 FOR IR = 1 TO NR
670 Y(IR,IV) = −X(IR,IV)
680 NEXT IR
690 T = LO(IV)
700 LO(IV) = −HI(IV)
710 HI(IV) = −T
720 AV(IV) = −AV(IV)
730 NEXT IV
740 REM
750 REM Calculate scores.
760 REM
770 FOR IV = 1 TO NV
780 K = (NR * LO(IV) − HI(IV))/(NR − 1)
790 FOR IR = 1 TO NR
800 Y(IR,IV) = Y(IR,IV) − K
810 SC(IR) = SC(IR)*Y(IR,IV)
820 NEXT IR
830 NEXT IV
840 REM
850 REM Sort from high to low index score
860 REM
870 FOR IR = 1 TO NR − 1
880 FOR JR = IR + 1 TO NR
890 IF SC(JR) < = SC(IR) THEN 1010
900 T = ID(IR)
910 ID(IR) = ID(JR)
920 ID(JR) = T
930 FOR IV = 1 TO NV
940 T = X(IR,IV)
950 X(IR,IV) = X(JR,IV)
960 X(JR,IV) = T
970 NEXT IV
980 T = SC(IR)
990 SC(IR) = SC(JR)
1000 SC(JR) = T
1010 NEXT JR
1020 NEXT IR
1030 REM
1040 REM Print data and scores in sorted order
1050 REM
1060 REM Standardize scores before printing
1070 REM
1080 SMIN = SC(NR)
1090 SMAX = SC(1)
1100 FOR IR = 1 TO NR
```

```
1110 SC(IR) = (SC(IR) − SMIN)/(SMAX − SMIN)*100
1120 NEXT IR
1130 PRINT "List data to printer in sorted order"
1140 LPRINT "Listing of data from file = "; FI$
1150 LPRINT USING " ### genotypes; ## variables "; NR,NV
1160 LPRINT
1170 LPRINT "Overall scores calculated by Elston's method"
1180 LPRINT " N.B. Scores standardized to 0—100 scale"
1190 LPRINT
1200 LPRINT "Identif.      Data";:LPRINT TAB(55),"Score"
1210 LPRINT
1220 FOR IR = 1 TO NR
1230 LPRINT USING "     ####      "; ID(IR);
1240 FOR IV = 1 TO NV
1250 LPRINT USING "     ####.###"; X(IR,IV);
1260 IF INT(IV/5) * 5 = IV THEN LPRINT:LPRINT TAB(10)
1270 NEXT IV
1280 LPRINT TAB(60);
1290 LPRINT USING " ####.#";SC(IR)
1300 NEXT IR
1310 CLOSE 1
1320 LPRINT   "= = = = = = = = = = = = = = = = = = = = = = = = = = =
     = = = = = = = = = = = = = = = = = ="
1330 PRINT
1340 END
```

APPENDIX 13:

PROGRAM FOR EVALUATING POTENTIAL PARENTS

PROGRAM NAME: PEDERSON.BAS

LANGUAGE: BASIC

PURPOSE:

This program compares the mid-parent values for certain types of crosses among all possible combinations of a given set of parents with a specified ideal, and identifies those crosses that most closely approach the ideal.

DESCRIPTION:

Following the suggestion by Pederson, crosses are compared to a specified ideal by calculating a weighted sum of squares of deviations between the ideal values and the average values of two or more parents. The weights used for each trait are the reciprocal of the squares of the "maximum preferred deviations" as specified by the user. It is assumed that genotypic effects are primarily additive so that average value of the parents, each weighted by their genetic contribution to the offspring population, will be a good indicator of the progeny mean. The closer the progeny mean is to the ideal value, the easier it should be to identify one or more progeny lines with suitable combinations of all traits considered in the parental selection process.

In many applied breeding programs, it is unlikely that more than two crossing cycles would be used for developing progeny populations. For this reason, this program evaluates the weighted deviation sums of squares only for single crosses and single backcrosses between two parents, and for three- and four-way crosses. The combination of parents that gives the lowest deviation sum of squares in each type of cross is identified.

USER INSTRUCTIONS:

The user must specify the file that contains the parental identification numbers and data, and must also specify the number of traits for each potential parent and the number of parents. The program reads the data from the input file and prints it on the screen. The user is then prompted to indicate the ideal value and the maximum preferred deviation for each trait.

Single crosses between all possible pairs of parents and the two possible backcrosses between all possible pairs of parents are first compared to the ideal. Then the three possible three-way crosses among all possible triplets of parents and the four-way crosses among all possible quadruplets of parents are evaluated. Output consists of a list of the parents showing the lowest deviation sum of squares in each class as well as the deviation sums of squares.

If the user wishes, the program could be easily modified to print the weighted deviation sum of squares for each cross as it is calculated. For example, to print the deviation sums of squares for each possible single cross, one would add a statement such as 875 PRINT I1,I2, SD before running the program.

As indicated in the discussion in Chapter 10, the number of crosses to be evaluated increases rapidly as the number of parents increases. With as few as 20 potential parents, the program evaluates a total of 8835 crosses.

PROGRAM LISTING:

```
10 REM Program to perform modified Pederson evaluation of
20 REM potential parents. Weighted squares of deviations
```

```
 30 REM between mid-parent and ideal calculated for all
 40 REM possible single, single backcross, three-way and
 50 REM four-way crosses. Weighting according to
 60 REM Pederson, D. G., A least-squares method for choosing
 70 REM the best relative proportions when intercrossing
 80 REM cultivars, Euphytica,30,153,1981.
 90 REM
100 REM Program reads identification number and data for
110 REM each potential parent from a file, prompts user
120 REM to specify ideal values and maximum preferred
130 REM deviations for each variable, evaluates crosses
140 REM and displays best cross in each category.
150 REM
160 REM Turn key definitions off and clear screen.
170 REM
180 KEY OFF:CLS
190 PRINT
200 PRINT" Parental evaluation by modification of Pederson's method"
210 PRINT "= = = = = = = = = = = = = = = = = = = = = = = = = = = = = =
    = = = = = = = = = = = = = = = = = = = = = = ="
220 PRINT
230 INPUT "What is name of file containing parental data "; FI$
240 OPEN FI$ FOR INPUT AS #1
250 PRINT
260 INPUT "How many traits measured on each parent "; NV
270 PRINT
280 INPUT "How many parents to be evaluated "; NP
290 PRINT
300 DIM LD(NP),X(NP,NV),ID(NV),MD(NV),SS(5,5)
310 REM
320 REM Set residual sums of squares to 10,000.
330 REM
340 FOR J = 1 TO 5
350 SS(J,1) = 10000
360 NEXT J
370 REM
380 REM Read and print identification numbers and data
390 REM
400 PRINT
410 PRINT "Listing of parental identification and data"
420 PRINT
430 FOR IP = 1 TO NP
440 INPUT #1,LD(IP)
450 PRINT USING " #### "; LD(IP);
460 FOR IV = 1 TO NV
470 INPUT #1,X(IP,IV)
480 PRINT USING " ####.##"; X(IP,IV);
490 NEXT IV
500 PRINT
510 NEXT IP
520 REM
```

```
530 REM Prompt for ideal values and maximum preferred deviations.
540 REM
550 FOR IV = 1 TO NV
560 PRINT USING "For variable ##, what is the ideal value "; IV;
570 INPUT ID(IV)
580 PRINT " What is the maximum preferred deviation ";
590 INPUT MD(IV)
600 PRINT
610 NEXT IV
620 REM
630 REM Start loop to look for best crosses in each category.
640 REM
650 CLS
660 PRINT " Looking for best crosses "
670 REM
680 REM Pick parent that is closest to ideal.
690 REM Print heading for parent numbers being evaluated.
700 REM
710 LOCATE 4,1:PRINT" P1 P2 P3 P4"
720 FOR I1 = 1 TO NP
730 SD = 0
740 FOR IV = 1 TO NV
750 SD = SD + ((ID(IV) − X(I1,IV))/MD(IV))^2
760 NEXT IV
770 IF SD > SS(1,1) THEN 830
780 SS(1,1) = SD
790 SS(1,2) = LD(I1)
800 REM
810 REM Pick best single cross.
820 REM
830 FOR I2 = I1 + 1 TO NP
840 SD = 0
850 FOR IV = 1 TO NV
860 SD = SD + ((ID(IV) − .5*X(I1,IV) − .5*X(I2,IV))/MD(IV))^2
870 NEXT IV
880 IF SD > SS(2,1) THEN 950
890 SS(2,1) = SD
900 SS(2,2) = LD(I1)
910 SS(2,3) = LD(I2)
920 REM
930 REM Pick best backcross (A × B) × B
940 REM
950 SD = 0
960 FOR IV = 1 TO NV
970 SD = SD + ((ID(IV) − .25*X(I1,IV) − .75*X(I2,IV))/MD(IV))^2
980 NEXT IV
990 IF SD > SS(3,1) THEN 1060
1000 SS(3,1) = SD
1010 SS(3,2) = LD(I1)
1020 SS(3,3) = LD(I2)
1030 REM
```

```
1040 REM Pick best backcross (A × B) × A
1050 REM
1060 SD = 0
1070 FOR IV = 1 TO NV
1080 SD = SD + ((ID(IV) − .75*X(I1,IV) − .25*X(I2,IV))/MD(IV))^2
1090 NEXT IV
1100 IF SD > SS(3,1) THEN 1140
1110 SS(3,1) = SD
1120 SS(3,2) = LD(I2)
1130 SS(3,3) = LD(I1)
1140 FOR I3 = I2 + 1 TO NP
1150 REM
1160 REM Pick best three-way cross (A × B) × C
1170 REM
1180 SD = 0
1190 FOR IV = 1 TO NV
1200 SD = SD + ((ID(IV) − .25*X(I1,IV) − .25*X(I2,IV) − .5*X(I3,IV))/MD(IV))^2
1210 NEXT IV
1220 IF SD > SS(4,1) THEN 1300
1230 SS(4,1) = SD
1240 SS(4,2) = LD(I1)
1250 SS(4,3) = LD(I2)
1260 SS(4,4) = LD(I3)
1270 REM
1280 REM Pick best three-way cross (A × C) × B
1290 REM
1300 SD = 0
1310 FOR IV = 1 TO NV
1320 SD = SD + ((ID(IV) − .25*X(I1,IV) − .5*X(I2,IV) − .25*X(I3,IV))/MD(IV))^2
1330 NEXT IV
1340 IF SD > SS(4,1) THEN 1420
1350 SS(4,1) = SD
1360 SS(4,2) = LD(I1)
1370 SS(4,3) = LD(I3)
1380 SS(4,4) = LD(I2)
1390 REM
1400 REM Pick best three-way cross A × (B × C)
1410 REM
1420 SD = 0
1430 FOR IV = 1 TO NV
1440 SD = SD + ((ID(IV) − .5*X(I1,IV) − .25*X(I2,IV) − .25*X(I3,IV))/MD(IV))^2
1450 NEXT IV
1460 IF SD > SS(4,1) THEN 1510
1470 SS(4,1) = SD
1480 SS(4,2) = LD(I2)
1490 SS(4,3) = LD(I3)
1500 SS(4,4) = LD(I1)
1510 FOR I4 = I3 + 1 TO NP
1520 REM
1530 REM Pick best four-way cross (A × B) × (C × D)
1540 REM
```

```
1550 SD = 0
1560 FOR IV = 1 TO NV
1570 SD = SD + ((ID(IV) − .25*(X(I1,IV) + X(I2,IV) + X(I3,IV) + X(I4,IV)))/MD(IV))^2
1580 NEXT IV
1590 IF SD > SS(5,1) THEN 1680
1600 SS(5,1) = SD
1610 SS(5,2) = LD(I1)
1620 SS(5,3) = LD(I2)
1630 SS(5,4) = LD(I3)
1640 SS(5,5) = LD(I4)
1650 REM
1660 REM Print parent numbers to show progress
1670 REM
1680 LOCATE 5,1:PRINT USING " ### ### ### ###";I1,I2,I3,I4
1690 NEXT I4
1700 NEXT I3
1710 NEXT I2
1720 NEXT I1
1730 REM
1740 REM Print results to screen.
1750 REM
1760 CLS
1770 PRINT
1780 PRINT USING "Parent ### is closest to ideal with deviations ss of
     #####.###";SS(1,2),SS(1,1)
1790 PRINT
1800 PRINT USING "The best single cross is ### X ### with deviation ss of
     #####.###";SS(2,2),SS(2,3),SS(2,1)
1810 PRINT
1820 PRINT USING "The best backcross is (### X ###) X ### with deviation ss of
     #####.###";SS(3,2),SS(3,3),SS(3,3),SS(3,1)
1830 PRINT
1840 PRINT USING "The best 3-way cross is (### X ###) X ### with deviation ss
     of #####.###";SS(4,2),SS(4,3),SS(4,4),SS(4,1)
1850 PRINT
1860 PRINT USING "The best 4-way cross is (### X ###) X (### X ###)
     with";SS(5,2),SS(5,3),SS(5,4),SS(5,5)
1870 PRINT TAB(36);
1880 PRINT USING "deviation ss of #####.###"; SS(5,1)
1890 PRINT "= = = = = = = = = = = = = = = = = = = = = = = = = = = = = = =
     = = = = = = = = = = = = = = = = = = = ="
1900 PRINT
1910 END
```

INDEX